WAGING WA

A PHILOSOPHICAL INTRODUCTION

WAGING WAR

A PHILOSOPHICAL
INTRODUCTION

IAN CLARK

CLARENDON PRESS · OXFORD

Oxford University Press, Walton Street, Oxford OX2 6DP
Oxford   New York   Toronto
Delhi   Bombay   Calcutta   Madras   Karachi
Kuala Lumpur   Singapore   Hong Kong   Tokyo
Nairobi   Dar es Salaam   Cape Town
Melbourne   Auckland   Madrid
and associated companies in
Berlin   Ibadan

Oxford is a trade mark of Oxford University Press

Published in the United States
by Oxford University Press Inc., New York

British Library Cataloguing in Publication Data
Clark, Ian
Waging war: a philosophical
introduction
1. War (Philosophy)
I. Title.
355'.001   U21.2
ISBN 0–19–827759–8 (Pbk)

Library of Congress Cataloging in Publication Data
Clark, Ian, 1949–
Waging war: a philosophical introduction / Ian Clark.
p.   cm.
Bibliography: p. Includes index.
1. War   2. War—Moral and ethical aspects.   I. Title.
U21.2.C52 1988   355'.02–dc19   87–30754
ISBN 0–19–827759–8 (Pbk)

3 5 7 9 10 8 6 4

Printed in Great Britain
on acid-free paper by
Biddles Ltd,
Guildford and King's Lynn

CONTENTS

INTRODUCTION

WAR is a supremely practical activity. The destruction of the battlefield, it would seem, is a universe removed from the sedate reflections of the philosopher. As such, war is more appropriately regarded as a realm of action than as a realm of abstract contemplation: if we seek to know about the nature of war, we should immerse ourselves in works of military history, not in works of moral or philosophical theory.

In these brief observations we find a statement of the immediate difficulties in approaching the subject of war, and the waging of it, from an essentially philosophical perspective. The first problem is one of resistance. To many minds the effort to reduce the death, pain, and suffering inflicted by war to a set of philosophical speculations and abstractions is quaint at best and obscene at worst. Hence, to the military pragmatist, philosophy of war is the last refuge of the academic scoundrel who claims to some knowledge about the subject of war—but at a very safe distance.

The theorist is twice cursed. Not only is he remote from the subject about which he writes but his labours are, inescapably, in vain. The gulf between practice and theory is such that whatever general guidelines might be devised for the conduct of war, in terms of idealized sets of principles, must inevitably break down when confronted with real situations. General principles tell us little about conduct in circumstances where competing principles apply or, indeed, in the grey areas where the general principles imperfectly capture the complexity of reality. Choices in war are, by their nature, hard choices which result in death and destruction whichever course of action is adopted. Those who saw virtue in the Second World War strategic bomber offensive against Germany did not do so *in vacuo* but as the lesser evil in comparison with the grisly land campaigns of the First World War. Bomber Command's Arthur Harris accordingly preferred an air strategy to 'morons volunteering to get hung on the wire and shot in the stomach in the mud of Flanders'.[1] From this point

---

[1] M. Smith, *British Air Strategy between the Wars* (Clarendon Press, 1984), 64.

of view, whatever philosophical reflection might tell us about the *essential* nature of warfare, it can have minimal impact on its actual conduct.

This suggests that there is little common ground between the practice of war and the activity of philosophical contemplation upon it. The two are worlds apart. This was to be the point at issue in the dispute between Paul Fussell and Michael Walzer over the ethics of the atomic bombings of Hiroshima and Nagasaki.[2] Fussell, on that occasion, espoused the view that 'experience' of war provided a critical perspective on the subject: 'the experience I'm talking about is that of having come to grips, face to face, with an enemy who designs your death.' When it comes to discussion of the legitimacy of acts of war, Fussell would argue, such experience concentrates the mind wonderfully and provides unique insight into the 'reality' of war. While he does not quite say that those without this experience are disqualified from the discussion, he comes mightily close to implying as much, as in his innuendo that J. K. Galbraith's interpretation of the bombing is invalidated by the safe position he held in 1945: 'I don't demand that he experience having his ass shot off,' Fussell commented, 'I just note that he didn't.'

If Fussell insists that having been close to the action provides a special vantage-point for looking at the subject, Walzer has argued against this that it may also bring one *too* close to the action. The perspective will thereby be distorted and essentials will be left out of the picture by the very closeness of the fighting. After all, war is more than the actuality of combat: participation in the latter offers, at best, a partial understanding of the whole. What we refer to as a state of war encompasses a complex of legal and political conditions, over and above the fighting that takes place.

Fussell is surely wrong on two counts. First, on general grounds, he errs in suggesting that moral judgement is tied to direct experience. This is not deemed to be a necessary qualification to participate in other forms of moral discourse and, indeed, such a personal perspective may be detrimental to objectivity. Secondly, he underestimates the extent to which the gulf can be bridged, in any case, by empathy and creative imagination. Is it not ironic that a book, widely praised as giving a brilliant portrait of the reality of *The Face of Battle*,[3]

---

[2] 'Hiroshima: A Soldier's View' and 'An Exchange on Hiroshima', *The New Republic*, 22, 29 Aug., 23 Sept. 1981.
[3] John Keegan's book by that name (Jonathan Cape, 1976).

should have opened with the disclaimer that the author had 'not been in a battle; not near one, nor heard one from afar, nor seen the aftermath'?

Understanding war, in its full philosophical richness, requires several orders of empathy beyond the simple capacity to visualize the horrific experiences of others. Judgements about war can scarcely avoid the delicate issues of generational responsibilities and obligations, both in terms of the causes of wars and of their long-term consequences. There is no necessary coincidence between the state's official decision-makers and those who suffer directly in its name, and some theoretical framework is required to integrate the two; in its absence, empathy breaks down with corrosive impact, as in the anger of war poetry:

> The young men of the world
> Are condemned to death.
> They have been called up to die
> For the crime of their fathers.[4]

The generational relationship need not be so fractious. It can exist also in the empathic obligation to pursue the cause of those who have died in its quest:

> Take up our quarrel with the foe:
> To you from failing hands we throw
> The torch; be yours to hold it high.
> If ye break faith with us who die
> We shall not sleep, though poppies grow
> in Flanders fields.[5]

A philosophical account of war must be sufficiently comprehensive to embrace fathers and sons, the living and the dead.

In any event the separation between philosophy and practice is much less than complete. Few military pragmatists, if pressed hard, are likely to take the position that there are no restrictions whatsoever upon the waging of war: military manuals the world over refute any such idea. Even those who subscribe to the notion that 'war is hell' do so in a relative, rather than an absolute, way. The experience of war is mightily different from the experience of peace but the transition from

---

[4] F. S. Flint, 'Lament', in J. Silkin (ed.), *First World War Poetry* (Penguin Books, 1979), 147.
[5] J. McCrae, 'In Flanders Fields', ibid. 85.

one to the other does not take us from order to total anarchy. Nardin makes the point well:

it does not follow from the fact that in war the normal order of society is disrupted that the state of war is one without order. The alternative to life according to one set of rules is not necessarily life without any rules at all, but rather life according to different rules.[6]

Once this crucial concession is made, we engage in a dialogue about where and why the lines of restraint are to be drawn and already the philosophy of war is begun.

It is not the objective of this work to offer a set of prescriptions for the conduct of war, nor to elaborate a single philosophical orientation towards the nature of war and the manner of waging it. Its purpose will be served if it acts as a guide to the multiplicity of complex issues which compete for attention when this subject is considered.

Beyond this limited goal, the intention of this book is to offer a tentative introduction to the manner in which the philosophy and practice of war might be integrated. Its point of departure is thus an absolute denial of the assumption that war can be 'practised' in separation from a theoretical understanding of its nature. In short, it is only by knowing what it is that we practise that any kind of framework for the discussion of the waging of war can be constructed at all. As Michael Oakeshott suggested in a more general context 'the so-called "practical" is not a certain kind of performance; it is conduct in respect of its acknowledgement of a practice'.[7] When we thus refer to the practice of war, we do not refer simply to the actions which people perform but to the context of choice and understanding in which certain acts of violence are recognized as acts of war.

While sympathizing, therefore, with a recent book which demonstrates the difficulties of distinguishing war from non-war, as regards differing forms of communal violence, and while accepting that the drawing of such boundaries is fraught with difficulty, the present work rejects the conclusion that 'it is not essential to be able to distinguish cleanly between war and near-war. The importance and the immediacy of the moral problems will remain constant.'[8] If the practical conduct of war is inextricably related to a practice of war, and this in turn to

[6] T. Nardin, *Law, Morality and the Relations of States* (Princeton University Press, 1983), 288.
[7] *On Human Conduct* (Oxford University Press, 1975), 57.
[8] N. Fotion and G. Elfstrom, *Military Ethics* (Routledge & Kegan Paul, 1986), 3.

acceptance of a common set of understandings, it follows surely that our moral assessments will vary in accordance with our degree of common recognition of war's nature and of its necessary features. To this extent, war is voluntaristic and we make of it what we will. The practice of war is related to the practice of peace. For some, peace can be defined negatively as the simple absence of war. Those who have written *The History of Peace* have generally written books about the elimination of the practices of war.[9] More interestingly, however, others have seen the proper end of war as the restoration of a better peace. General Sherman, however inappropriately, is commemorated with the epitaph '[T]he legitimate object of war is a more perfect peace'.[10] Accordingly, it is also the task of a philosophy of war to depict it integrally within the landscape of peace.

War has come to be regarded as an adjunct of the modern state system and a major part of the intellectual framework which surrounds it is provided by the heritage of ideas about the nature of international society, and about the place of individuals and states within it. It is apposite, therefore, to note that many contemporary issues in the philosophy of war are reflected in related debates about the wider realm of international relations. If a central schism in Western thought about international relations has been that between the respective obligations of 'men and citizens',[11] then recent theoretical appreciations of the nature of international society have clearly demonstrated the practical import of these competing conceptions. It is not too much to say that the practices of the state system have been critically examined in recent years as to their consistency with 'individual' and 'human' values of one kind or another. The place of human rights in the conduct of international relations, the legitimacy of intervention in the affairs of other states, and the nature, if any, of obligations to a more egalitarian distribution of the global product—all these matters have received renewed scrutiny and demonstrate the impact of changing intellectual frameworks upon the practices of statecraft.[12]

Recent analyses of war have been subject to similar intellectual

[9] See the book of that name by A. C. F. Beales (Bell & Sons, 1931).
[10] J. F. C. Fuller, *The Conduct of War, 1789–1961* (Methuen, 1961), 111.
[11] This theme is developed in A. Linklater, *Men and Citizens in the Theory of International Relations* (Macmillan, 1982).
[12] See e.g. C. Beitz, *Political Theory and International Relations* (Princeton University Press, 1979); R. J. Vincent, *Human Rights and International Relations* (Cambridge University Press, 1986); C. Beitz *et al.* (eds.), *International Ethics* (Princeton University Press, 1985).

influences. If war is deemed to be an institution of international society, its nature will change in accordance with changing conceptions of that societal framework. International society is no longer generally viewed as exclusively formed by corporate states: it is the milieu in which the world of states intermingles with the world of people. Hence, such general issues about duty and obligation which emerge from revised conceptions of international (or world) society cannot but impinge upon the specific realm of warfare. This is manifested in recent discussions about individual responsibility in war, the relationship between state and individual as a basis for discrimination in targets of warfare, and the philosophical bases of the rules of war generally. Put briefly, it might be said that the central issue is that of reconciling theories of human rights, and associated ideas of 'private' morality, with the corporate nature of warfare and the 'public' demands of the state-at-war.[13]

Much of the interest in the philosophy of war centres upon the dichotomy between 'public' and 'private'. Not least is this so because of the augmented public capacity to do harm. As Nagel has complained 'the great modern crimes are public crimes . . . the growth of political power has introduced a scale of massacre and despoliation that makes the efforts of private criminals, pirates, and bandits seem truly modest.'[14] The ramifications spread beyond this fact alone. How is it that war has been able to mobilize resources on this public scale? As a historian of Renaissance warfare has proclaimed 'the central mystery of politicized conflict is not why wars took place but how enough men could be found to fight in them'.[15] A resolution of this problem requires detailed historical knowledge of military administration and recruitment, of finance, and of the changing face of battle itself. It also demands a proper understanding of the intellectual underpinnings of the public state-at-war. This is the more so since the actual military art has itself become increasingly depersonalized. Heroic individual virtues are devalued in over-the-horizon warfare. As a historian of the technology of warfare has argued 'the technology of modern war, indeed, excludes almost all the elements of muscular heroism and simple brute ferocity that once found expression in hand-to-hand

---

[13] This is certainly one theme of M. Walzer, *Just and Unjust Wars* (Basic Books, 1977).
[14] T. Nagel, 'Ruthlessness in Public Life', in S. Hampshire (ed.), *Public and Private Morality* (Cambridge University Press, 1978), 75.
[15] J. R. Hale, *War and Society in Renaissance Europe* (Fontana, 1985), 45.

combat'.[16] In short, even the practice of warfare has lost some of its rudimentary private characteristics.

In accordance with this perspective the following chapters serve as an introduction to some of the issues that are raised by looking at the philosophy and practice of war, not as polar opposites, but as proximate realms of activity, each influencing the character of the other. This is not to say that the soldier, in the heat of battle, is engaged in philosophy but only that the nature of his activity has a meaning which derives from the framework of ideas surrounding it. Various symbolic acts of war—its formal declaration, treatment of prisoners and wounded, the niceties of battle itself, and acts of surrender such as the raising of hands—all derive their practical effect from a common appreciation of the nature of the activity in which the belligerents are engaged.

Accordingly the approach to this book will be to argue that there is an intimate interrelationship between how war is thought about and how it is waged. The first three chapters of the book can, therefore, be seen as deductive in method. They introduce various concepts of war and deduce certain means for the prosecution of war from these initial conceptions. In this way it can be shown that certain restraints on the waging of war make sense only within the context of a conception of war that is compatible with them. For instance, there is no point in making appeal to 'fair play' in war if your initial conception of war is one that makes no allowance for the notion of rules. As a specific manifestation of this, there can be no objection to bombing of civilians in cities if your concept of war is open-ended and contains no rules about the nature of targets. Chapter 1 will canvass this particular issue in some detail.

Chapters 2 and 3 will attempt to establish two major traditions about the waging of war, each of which argues for restraints in so doing, but each of which presents a substantially different argument for the kind of restraint it advocates. The main themes in, and the development of, the just war tradition will be discussed in Chapter 2. Chapter 3 will seek to present an alternative tradition, that of limited war, which, even when coming to similar conclusions for the practical conduct of war, reaches them by a different process of reasoning. In each case it will be contended that the practice of war is grounded in a distinctive conception of the nature of war itself. Although presented as two

[16] W. H. McNeill, *The Pursuit of Power* (Blackwell, 1983) p. viii.

separate traditions, just and limited war will also be compared and the
area of overlap between the two examined.

In Chapters 4 to 6 there is a shift of focus. Rather than seek to
deduce certain norms for the waging of war from an initial
philosophical orientation, the analysis will canvass inductively the
various discrete elements in terms of which war can be waged—issues
concerning the nature, extent, and means of the combat itself—and
discuss the practical issues at stake in implementing general principles
of warfare. Deductively, for instance, we might argue for the rights of
non-combatants from a definition of war as a contest between military
forces. Inductively, we might arrive at conclusions about what it is
permissible to do in warfare on the basis of various principles grounded
in theological premisses or theories of human rights. In terms of the
latter, killing innocent civilians is prohibited as a general principle of
ethics and not simply as a consequence of a particular conception of
the nature of warfare. In the former, the practice of war is deduced
directly from a concept of war that contains within it certain gross
restrictions on its prosecution.

These chapters will review the major issues of contemporary
warfare such as discrimination, proportionality, the nature of permissible
weapons, the idea of war crimes, the problems associated with nuclear
strategies, and the specific questions presented by the nature of
deterrence. In other words, in considering deterrence as a means of
preserving peace, we are compelled to pose the question whether
principles elaborated for the waging of war are relevant in the context
of strategies for the waging of peace: is there a practice of peace which
requires us to adopt different rules of conduct from those pertaining in
war or, on the contrary, might it be said that the practice of war only
has recognition within the greater goal of peace and derives its
meaning from that source?

The book will have three principal areas of concentration. The first
concerns the nature of war itself and how the central idea has
developed in the context of changing social, political, and technological
environments. The second is devoted to the elaboration of more
specific codes for the waging of war and explains how these have
variously evolved under the pressure of similar developments. Thirdly,
the book will analyse the problems of implementation and the means
by which general and abstract codes of conduct can be applied to the
practicalities of war. In examining this set of problems, the book will
focus on the ambivalence of 'soundness' in relation to the rules of

war—whether their soundness resides in philosophical strength and consistency or whether the major virtue of the rules of war is to be found in salience and observability. This is the real meeting ground of the theory and practice of war. Without philosophical substance and guidance, the codes of war are likely to be arbitrary and, in the nature of things, morally repugnant. At the opposite extreme, without any prospect of observance, the most rigorous philosophical systems have no material impact and will remain as abstract and unattainable ideals. It is at the point where these two forces come together—general philosophical orientations and the face of battle—that the real challenges of the philosophy of war are to be discovered.

I

WAR: CONCEPT AND CONDUCT

IN this book philosophy is presented as a means to a fuller understanding of the practice of war and not as a radical alternative to that practice. The entire argument is based on the contention that we cannot comprehend the manner of waging war without some wider framework of ideas within which acts of war, both of commission and omission, have meaning. Accordingly, the task of this chapter is to establish the general context of the discussion by demonstrating how certain conceptions of war entail, necessarily, certain modes of prosecution and how we can only come to terms with the means of waging war by locating them in their distinctive philosophical setting. This is simply to adapt the procedure of Machiavelli, of whom it has been said that all his military views were 'based on Machiavelli's concept of war and derive from it. . . . War . . . must end in a decision, and a battle was the best method of reaching a quick decision.'[1]

Just as the racing car driver handles his car differently from the family tourist because to be engaged in racing is, essentially, to be engaged in a different practice from that of touring, so the soldier, fighting in a war, practises a distinctive form of violence. A specific version of this argument has been advanced elsewhere:

What I should like to be able to do is to derive the morally defined limits from the very concept of war itself, so that if these are broached we no longer have war but slaughter, and whether there are moral limits to slaughter hardly can arise since slaughter is the other side of moral limits already.[2]

Danto concentrates upon one particular concept of war and derives from it certain rules for its conduct: this is not the only possible concept but, for the moment, we can concur in his general procedure, if not necessarily in his substantive conclusions.

This is not to claim that the style of war is determined by philosophers. The actual conduct of war is a product of a vastly

[1] F. Gilbert, 'Machiavelli: The Renaissance of the Art of War', in E. M. Earle (ed.), *Makers of Modern Strategy* (Princeton University Press, 1943), 22.
[2] A. C. Danto, 'On Moral Codes and Modern War', *Social Research*, Spring 1978, 180.

complex interplay of forces including political, social, economic, military, cultural, and technological factors. For this reason Michael Howard was able to discuss *War in European History*[3] in terms of the prevailing political arrangements for war at various historical periods. But, in saying this, we are not denying the role of ideas in shaping the manner in which our wars are fought. It is, after all, fundamental ideas about political and economic organization—such as the right of the prince to pursue public war, the legitimacy of the principle of national self-determination, and the need for a state structure to manage the technological scale of war—which have contributed to the resources available for the waging of war and which have influenced the ends for which it is fought. While it is true, therefore, that war is a supremely practical activity when once undertaken, we must not blind ourselves to the clash of ideas which has organized the warring parties, allowed them to mobilize their respective forces, and provided the intellectual imagery which serves as the axis of contention. This is not to say that wars are necessarily fought over ideas; merely that ideas inevitably play their part in the practical arrangements for war. Few developments have so profoundly affected the character and scale of contemporary warfare as the raising of mass conscript armies but this would not have been possible without underlying theoretical assumptions about the relationship between national state and citizen, or indeed, as Iran's recent experience reminds us, between the individual and the theocratic state.

The idea of war itself is a major factor in the way in which it is waged. We can, therefore, agree wholeheartedly with the remarks of Geoffrey Best in his introduction to the series of histories on war and society:

The idea of war is of itself a matter of giant historical importance: how at particular epochs and in particular societies it is diffused, articulated, coloured and connected. Only by way of that matrix of beliefs about God and man, nature and society, can come full understanding of the causes and courses of wars that have happened.[4]

What Best's remark impresses upon us is that, although we speak of war in human history as if it were a continuous and unchanging social institution, the context of ideas in which war has occurred has changed dramatically over time. Wars have been variously understood as an

---

[3] Oxford University Press, 1976.
[4] Series Editor's Preface to Fontana History of War and European Society.

affliction imposed by the gods, as a test of individual will and courage under supreme emergency, as a mere instrumentality of the dynastic ruler, and as the ultimate expression of the national essence. It is questionable whether one practice of war underlies these varying intellectual frameworks. In short, the attempt to understand war in history is unavoidably, albeit not exclusively, an excursion into the history of ideas.

War and the State

The intimate relationship between concept and conduct of war is nowhere clearer than in the view of war as something which happens between states. The theory of the state system and the practice of international law have long been predicated on the notion that war is a relationship between states and that this distinguishes war from other forms of violence. We find such a definition of war advanced by Plato in his *Republic*: 'It seems to me that war and civil strife differ in nature as they do in name, according to the two spheres in which disputes may arise: at home or abroad. . . . War means fighting with a foreign enemy; when the enemy is of the same kindred, we call it civil strife.'

What is intriguing about Plato's formulation is that, having elaborated his concept of war, he proceeds directly to deduce a code of conduct from it: 'Observe then, that, in what is commonly known as civil strife . . . it is thought an abominable outrage for either party to ravage the lands or burn the houses of the other.' But what is impermissible in civil strife is clearly acceptable in waging a war with a foreign enemy. In other words, the more permissive code of conduct which applies in war is but a natural deduction from the nature of the activity itself. War occurs between states *by definition* and the restraints which operate in the civil sphere do not operate in relations between states. Plato's distinction between external (inter-state) war and internal (civil) war has been a profound and enduring one. Ironically, however, the historical record has if anything seen an inversion of Plato's code of practice inasmuch as the conduct of external wars has been more effectively regulated and restrained, by the mechanisms of the international system, than has the conduct of civil wars which take place in a fractured or vacuous political milieu. Indeed, if anything, the prize of accreditation to the international community, via international recognition of domestic legitimacy, has raised the stakes of civil strife and contributed to the impassioned, and cruel, nature of such struggle.

The emergence and development of the modern state has imposed its imprint upon the practice of war in a variety of ways. It might be useful at the outset to argue that this has occurred in two discrete phases. Initially the state brought a degree of regulation to the practice of war that was to be so assertive as to lead to the concept of war being purloined by the doctrine of the state itself. However, in the aftermath, the nature of the state has changed dramatically in the past several hundred years and much of the changing style of warfare has been derivative from the evolution of the theory and practice of statecraft. In the long history of political theory, the state and war are regarded as being closely interwoven. Indeed, they are seen as serving a reciprocal function. On the one hand, there are the many theories which discover the origins of war in the very nature of the state. This argument adopts two broad forms. The first is the generalized proposition that war derives from the foundation of the political state itself. To the extent that the state is a political construct to overcome the ills of the 'state of nature', the consequent anarchical international system is simply the external price we have to pay for the state's imposition of domestic order: of these various costs, war is but the highest. Walzer captures the argument succinctly in his observation that 'the corollary of the King's Peace, thus established, was the king's war'.[5] This is no more than to say that it is the very endeavour to pacify the territory of the individual state by bringing it under the rule of sovereignty which creates the second-level problem of a state of nature between the sovereigns themselves. Thus runs the classical Hobbesian account. 'By the beginning of the eighteenth century', Howard remarks, 'political thinkers in general thus saw war as a necessary evil to keep in check yet greater evils.'[6] Indeed, as the century wore on, this was to become less of a justification of war and more the basis of a critique of society.

Alternatively the argument has taken a more particularist form: war derives not from the nature of *the* state but from the nature of *some* states. In brief, origins of war have been located in the political shortcomings of certain kinds of 'defective' states. From this perspective, autocratic, capitalist, and communist regimes have, at one time or another, been claimed to possess innate proclivities towards war.[7]

[5] M. Walzer, *The Revolution of the Saints* (Harvard University Press, 1965), 274.
[6] *War and the Liberal Conscience* (Temple Smith, 1978), 21.
[7] This is K. Waltz's second image. See *Man, the State and War* (Columbia University Press, 1959).

According to this image, war is a product of the old order in pre-1914 Europe, or of a capitalist driven diplomacy or, more recently, the consequence of the behaviour of 'evil empires'. Even democracy has not escaped such an indictment, as in Montesquieu's paradox that, although democratic nations are peaceful, their armies seek war: 'of all armies, those most ardently desirous of war are democratic armies and, of all nations, those most fond of peace are democratic nations.'

Approached from the opposite end, the reciprocal relationship is formed by the view that the origins of the state can be found in war. According to this conception, the state is, in origin, a security community that emerged and developed in response to the demands of warfare. The circularity of state and war is thereby complete.

This view finds expression in Bodin's claim that 'reason and commonsense alike point to the conclusion that the origin and foundation of commonwealth was in force and violence' and in Machiavelli's contention that the foundation of all states is in 'good arms'. The revolution in war and the revolution in statecraft thereby coincide. Machiavelli was therefore associated with a new view of warfare precisely because he located it in a new conception of political life in which 'the pursuit of power was made a matter of permanent, systematic struggle and accumulation'.[8] What we are invited to accept by these descriptions is a narrative history of the emergence of modern European states by dint of war. It was by successfully asserting their claim to be able to prosecute war that the European monarchs pressed the reality of a claim to stately autonomy *vis-à-vis* the lingering pretensions of corporate European Christendom. At the same time the European monarchs monopolized the claim to wage war *vis-à-vis* the feudal nobility, thereby eliminating 'private' forms of conflict between baronial rivals. The capability to wage war thus became the most important badge of statehood. As Howard has commented, 'by the end of the 16th century the men, as it were, had become separated from the boys: it was clear which princes were capable of waging war on their own account and which were not.'[9] The *locus classicus* depicting the doctrinal foundations of this transformation is, of course, the work of Machiavelli:

it is Machiavelli's approach to the problem of war that most vividly sets him off from his contemporaries, in his realization that in the European world to come

⁸ Walzer, *Revolution*, p. 273.
⁹ Howard, *War in European History*, p. 22.

only mastery of the techniques of war would lead to mastery of the techniques of power in society.[10]

This reciprocal relationship is deemed by some historians to have been operative in more practical ways as well. In terms of historical evolution, changes in the nature of the state are believed to have shaped the practice of war and changes in the nature of war are believed to have shaped the practice of the state. Tilly demonstrates this reciprocal interaction in analysing the chain linking expansion of land armies, new efforts to extract resources, bureaucratic and administrative innovation to make this state extraction more efficient, popular resistance, renewed coercion, and, finally, the consequent requirement for military means. 'The formation of standing armies', he concludes, 'provided the largest single incentive to extraction and the largest single means of state coercion over the long run of European state-making.'[11] Indeed, it is this acceptance of the autonomy of the military–strategic dimension in state formation and development that has served as a criticism of the 'world systems' literature[12] and which is now receiving increasing recognition from sociologists, whose erstwhile tendency had been to explore the theory of the state in purely socio-economic terms.[13]

What the state has given to the practice of war is its scale (the state has been politically a most efficient means of mobilizing military resources, primarily personnel), bureaucratic structure (required to administer and deploy large standing or conscript armies), technology (the state alone could deploy the resources required for a Manhattan Project), and professionalization of the armed forces (a necessary adjunct of bureaucratic control and of political purpose). This last was a crucial development, directly instrumental in translating a concept of war into an actualized practice. McNeill writes as follows of the impact of the professionalization of warfare:

As a handful of sovereign rulers monopolized organized violence and bureaucratized its management in Europe, war became, as never before, the sport of kings. Since the sport had to be paid for by taxation, it seemed wise to leave the productive, taxpaying classes undisturbed. . . . For soldiers to

[10] R. Nisbet, *The Social Philosophers* (Paladin, 1976), 73.
[11] C. Tilly (ed.), *The Formation of National States in Western Europe* (Princeton University Press, 1975), 73.
[12] See e.g. A. Zolberg, 'Origins of the Modern World System: A Missing Link', *World Politics*, Jan. 1981.
[13] Most recently, A. Giddens, *The Nation-State and Violence* (Polity Press, 1985).

interfere with their activities was to endanger the goose that laid the golden eggs.[14]

In return, the demands of war have stimulated the mobilization of social and economic resources for civil ends, have led to the periodic reform, or revolution, of the state apparatus (not uncommonly in the aftermath of incompetence demonstrated in war itself, as in Russia in 1856, 1905, and 1917), and have had a catalytic effect upon social and political philosophies. At the same time theorists have rationalized the augmented role assumed by the state in the management of war in terms of its role in peace (management of the economy and the 'interventionist' state more generally), or have sought for programmes to vindicate the sacrifice endured by populations in the course of war (homes fit for heroes and welfarism as an incidental by-product of the two world wars).

In short, we can firmly relate the practice of the state to the practice of war in terms of their mutual development. For the present purpose the claim is both narrower in scope and more fundamental in consequence. It is that the steady flow of the river of state practice across the plains of war has left a substantial sedimentary deposit upon our philosophical conceptions of warfare and, by association, upon our attitudes towards the practices of war. At this stage, it will suffice to enumerate some of the ideas that have germinated in this rich, alluvial soil:

1. Inasmuch as war has predominantly, and certainly from the seventeenth to the mid-twentieth centuries, been defined as a relationship between states, there has been a philosophical disjunction between those varieties of violence that we call war and those other varieties of violence (civil war, national liberation struggles, transnational terrorism) which are beyond the metaphorical pale. Against the efforts of both the sociologists and some contemporary schools of peace research to systematize the study of human violence in all its rich profusion and to break down the barriers between the seemingly structured forms of political violence and its unstructured and anomic forms, mainstream Western political and international relations theorists have erected an impressive array of intellectual barricades. It is war's character as being national, rational, and instrumental, each

---

[14] W. H. McNeill, *The Pursuit of Power* (Blackwell, 1983), 161.

defined in association with state interest and values, that sets the violence of war apart from its other myriad embodiments.[15]

The international lawyers have operated dutifully as the handmaidens of political theory and practice. Around the conception of the state's unique qualification to engage in organized political violence, they have constructed the legal trappings of this doctrine and have laid down a code of legal conduct for the prosecution of war: by definition, however, and until relatively recently, other forms of armed conflict fell outside the purview of the laws of war. Traditionally, therefore, in terms of conduct, the rules for waging war have applied in *war* but not in other forms of violence. Already, we have an explicit link between concept and conduct.

2. It follows also that, if war is regarded as a relationship between states, it makes no sense to speak of it as a kind of primordial, natural instinct for violence. It is true that the state may simply be made in humanity's image, but it remains equally true *by definition* that to conceptualize war as something which can occur only between states is, of necessity, to present war as a social or political institution created on a corporate and non-personalized level. It is at this point that Fussell's insistence upon the worm's eye view of the individual combatant, in fear of life, is potentially most distorting. Although expressed in an obviously real sense as a direct relationship between individuals, war as combat is much more than this. While it is natural that the soldier will regard as the enemy the individual who is firing a machine gun at him, the notion of war as an inter-state activity should serve to depersonalize this relationship. Even if it is conceded that the state is moved by its own 'passions', this has to be understood as metaphor. The conscript in the opposing trench is at the very least relegated to the role of representative of the enemy: this makes his bullets no less deadly but it drives home the point that the ultimate purpose of the soldier's activity is not to bring about his antagonist's death, even when this may be an unavoidable means to the ends of war.

3. The theorist's task is by no means done when he has established that war is a form of violence that takes place between states. At the very least he must proceed to give some account of the nature of the state *qua* belligerent. What is the state when it fights? Is this to be understood as a social reality? Or is it merely employed as a convenient

---

[15] See e.g. A. Rapoport's critical introduction to C. von Clausewitz, *On War* (Pelican ed., Penguin Books, 1968).

symbol—a corporate myth—to refer to a much more complex reality? If, as some ideologies would have us believe, the state is a committee of its ruling class, how are we to distribute any moral obligation to die in its name? This already suggests enough to indicate that the practice of war is itself tied up with a sociological analysis of the state: the conduct of war has both been related to the historical development of state institutions and, as a matter of moral theory, must be related to some wider understanding of the individual's location within, or apart from, this institution. At this point the conventional distinction between state and society becomes a crucial one. If there is a demarcation between the two, as most liberal political theories would suggest, it is possible to build upon this distinction certain rules for the waging of war. If war is between states, and the state is separable in theory as well as in practice from society, it can be argued that this conception of war leads directly to a principle of discrimination in the conduct of war since the targets of war are, presumably, the rival *states* but by no means the entire populations of the belligerents.

To this extent, it should become readily apparent that stipulating the proper targets of war is itself an exercise in political philosophy dependent upon our overarching conception of war, our view of the state, and of its relationship to its own citizens. A proper understanding of the conduct of war requires us to come to terms with the multiplicity of philosophical problems generated by these particular issues.

Concepts of War

An investigation into the nature of war becomes a concrete task of drawing meaningful boundaries within the general realm of violence and consequent boundaries to demarcate the means by which this activity should be pursued. It is with the concept of war that the analysis must commence. Central to this discussion is the validity of distinguishing war from other political relations within the international system and hence the attempt to draw a sharp boundary between war and peace.

At one end of the spectrum we have those analysts who posit a radical separation between 'war' activities and 'peace' activities. Midlarsky might be taken as representative of this school:

We can understand war as a failure of normal power (political) relations, such that force (coercion), in the form of political violence, results. War is, then, not the 'continuation of political relations' but their termination in the onset of extreme coercion. Rather than a continuous political process, there occurs a discrete change from power to force.[16]

What has called this approach into doubt has been those characteristics of the 'diplomacy of violence' which have featured so prominently in post-1945 international relations, such as to evoke the term Cold War to account for a period of peace in which the violence of war seems to be present, even if latent. This incorporates a wider conception of war, as in the Hobbesian diagnosis of a 'known disposition thereto', regardless of whether or not overt hostilities are taking place.

It has, therefore, become common to question the separation between war and peace when threatened violence has become a conspicuous element in the very structure of peace. Stressing the role of technology in welding the two realms, Pearton thus argues that 'the polarity of "peace" and "war" has ceased to be axiomatic. Preparedness for war . . . is a continuous activity, reaching into all aspects of society and eroding, even nullifying, conventional distinctions about the "civil" and the "military" spheres of life.'[17]

So addressed, this development might appear a speculative one. It is, however, given tangible support by the state practice of no longer formally recognizing the condition of war, even when hostilities are manifestly taking place. Whereas previously it has always been essential that the state of war be pronounced for the laws of war to operate, current practice shies away from recognition of war, both by the adoption of euphemisms and by failure to declare that a state of war exists.[18]

How war is to be waged depends fundamentally upon what war is conceived to be. Historically, conceptions have changed, each reflecting its own distinctive historical circumstances. It would be impossible to provide a comprehensive survey of the variety of concepts of war and, accordingly, the following selection will be discussed for purposes of illustration to demonstrate the general theme that there is a crucial link between perceptions of the nature of war and the kind of conduct which it invites.

War as instinctive violence Within this frame of reference, war is

---

[16] M. I. Midlarsky, *On War: Political Violence in the International System* (Free Press, 1975), 1–2.

[17] M. Pearton, *The Knowledgeable State* (Burnett Books, 1982), 11.

[18] See S. Bailey, *How Wars End*, i (Clarendon Press, 1982).

understood simply as a manifestation of human nature and its propensity to violence. Since the violence of war is not a 'goal-oriented' activity, even though it may express itself in pursuit of a variety of secondary objectives, the nature and the extent of the violence is not controlled by its relationship to a specific end. In such a framework there can be no intrinsic measure of appropriate or excessive levels of force since the violence is its own end. It is difficult in this case to deduce any rules for the conduct of war from the initial concept and, accordingly, this concept is the exception.

At best this view of war allows for any manner of institutional restraint upon the death and destruction of war both by trying to limit the resort to it and by seeking to place limits upon the degree of violence employed. Such efforts must remain as no more than extrinsic palliatives: they do not derive from the purpose of war, nor can they hope to remedy the basic problem. They are no more intimately related to the nature, or cure, of leprosy than is the leper's bell to his disease. It should, therefore, be apparent that, in such a relationship, the conduct of war is a construct set up in opposition to the concept of war and not a set of practices logically deduced from it. The need to contain the ravages of such human instincts is established by other social values and purposes, divorced from the activity of war itself.

War as divination or legal trial It has been common to regard war as a means of achieving a divine or judicial verdict. Success in battle is thereby regarded as a tangible demonstration of divine support for the venture. The Romans took the precaution of taking the auguries before the fray whereas medieval armies took the field, content to await divine judgement in the battle itself; but in both cases the practice of war was related to a divine cosmology and to the human search for, or desire to capitalize temporally from, godly wishes. Likewise, trial by battle is equivalent to other judicial proceedings and is a means of arriving at a legal 'verdict'. The side which is successful in battle has had its legal suit vindicated.

If this be the purpose of the fray, it follows naturally that the war be waged with certain rules of equal advantage observed. Just as the various claimants in a judicial proceeding have a right to a fair trial, so in war steps should be taken to ensure that the parties have a fair chance of securing an appropriate verdict. This leads inevitably to a highly formalized style of battle in which 'cheating', by seeking to take undue advantage of the opponent, is proscribed. Thus the historian of the medieval laws of war tells us that when battles were an appeal to the judgement of God, 'every possible precaution had to be taken to thwart any attempt on the part of wicked men to weight the scales of divine justice'.[19] In this case, certain procedural rules for the conduct of war can be deduced from the essential nature of war itself.

War as disease The view of war as a pathological condition of the international body politic generates a distinctive normative orientation towards war. Its philosophical inclination is towards cure, even while allowing for the treatment of the symptoms. Consequently, this conception of war is more likely to be associated with programmes for the eradication of war than with programmes for restraining its conduct. To the extent that it takes account of the latter problem, it is with the intention of limiting the spread of infection and making the patient suffer as little as possible. However, the former is the primary objective (doctors are not content to alleviate symptoms where they can produce cures) and the latter a secondary palliative pending the attainment of the former.

A contrast can be drawn here with those closely related concepts of war as some kind of natural cataclysm or disaster, such as an earthquake or cyclone. From these perspectives, the approach is avowedly limitationist rather than abolitionist inasmuch as such natural disasters cannot be cured or abolished but measures can none the less be taken to minimize the extent of damage which they inflict.

War and social change This is a catch-all category. It includes two broad approaches—that war is a measure of social development and, alternatively, that war is a necessary means to social development. In terms of the former, the prevalence of war is attributable to social atavisms and will, in turn, be remedied by social change. In the theories of Comte, as of Schumpeter, war is associated with primitive social elements, and social and economic development, mostly to be brought about by industrialization, is a necessary prerequisite of the harnessing of war.

Such an attitude has bred a degree of complacency. As Buchan was to complain, the belief that supercession of agrarian, by industrial, society would suppress war contributed to a naïve faith in the automatic link between social progress and the decline of war as a social institution.[20]

[19] M. H. Keen, *The Laws of War in the Late Middle Ages* (Routledge & Kegan Paul, 1965), 130.   [20] A. Buchan, *War in Modern Society* (Collins, 1966), 1.

Reversing the sequence, a venerable tradition of social thought has regarded war as an instrument of social change, either by the maintenance of vital human values, or as a catalyst of social revolution.

War as a political instrument of the state This is the predominant political philosophy of war. Its point of departure is that war is undertaken for political ends and that, by implication, the means adopted for their attainment should be such as to further, and not contradict, those ends.

The extent to which such a concept of war lends itself to a philosophy of limitation in its conduct will be the subject of extensive analysis below. For the moment, it is sufficient to observe that this conception of war is related to conduct both in terms of the degree, and the nature, of the military force employed. There is at least an implicit suggestion that the degree of force be proportional to the objective being sought and, more generally, an admonition that the means not be counterproductive in terms of the stated ends. The relationship, at this level, is purely formal. It says no more than that, viewed as an instrumentality, the means and the ends of war are interrelated. In this form it tells us nothing about the conduct of war pursued for 'total' ends. Is the conception completely permissive? Nor does it tell us much about pragmatic calculations of margins of safety. How much 'extra' force can be allowed to ensure that objectives are attained?

War as regulator of the international system This conception comes from the same intellectual stable as the previous one but differs from it on account of the entity which has its interests served by resort to war. Whereas in the Clausewitzian paradigm, war is the instrument of the national state and is the means of securing its political objectives, this second political philosophy of war makes war the servant, not of the individual state, but of the international system itself. Wars are undertaken to preserve the integrity of the existing state system. Most notably such a conception is associated with theories of the balance of power which postulate war as the ultimate means of preserving or restoring a balance of power in the face of a challenge by a would-be hegemonic power. As Hedley Bull has written, of what he termed a Grotian conception of international society, 'war . . . derives its legitimacy from the service it renders to international society as a whole; the king or people going to war to redress an injury received are entitled to regard themselves as the instruments of a general purpose.'[21]

The intellectual proximity to the Clausewitzian concept is self-evident in two respects. First, it is our historical experience that the individual states themselves have decided where the best interests of international society lie, either by positive claim or by default, and have competed with each other to be the sacred trustee of that community's interests. Secondly, however, the idea of war to preserve the state system is but half an argument. It has scarcely been regarded by statesmen as an end in itself but more commonly as a means to state purpose: we come full circle in recognizing that, finally, the preservation of the state system is not an end in itself but a means to ensuring the security and survival of the individual state.

What the shift in the level of analysis from the state to the state system does do, however, is to suggest another measure of appropriateness in the conduct of war. If war is to serve as the regulator of the international system, it must be conducive and proportional to that end. This becomes of some importance in the discussion about nuclear weapons. It is one thing to say that an individual state might be able to resort to nuclear weapons in a politically instrumental way: it is another to suggest that war, entailing the widespread use of nuclear weapons, can continue to be regarded as a viable regulator of the international system. In short, certain means of war might well be disallowed if war is to fulfil its own stipulated functions.

As against this, however, we have the current idea of the nuclear deterrent as the latent regulator of the international system. The balance of power, and the survival of states within it, can be guaranteed by the fear of the great deterrent. Is this the final apotheosis of war? Does it mean that war has been abolished? Or does it mean that peace has been abolished? These questions will be explored below in the discussion of the relationship between deterrence, war, and its legitimate conduct.

Concepts and the Nature of Rules

What the above discussion has been attempting to address is the problem of deducing certain kinds of rules for the conduct of war from

---

[21] H. Bull, 'Grotian Conception of International Society', in H. Butterfield and M. Wight (eds.), *Diplomatic Investigations* (George Allen & Unwin, 1966), 56.

the very nature of the enterprise itself. The enquiry can advance by trying to answer three interrelated questions:

1. Is war a rule-governed activity at all?
2. If it is, what is the nature of the rules which govern it?
3. Does the value in having rules of war reside in the fact that there are rules, or does it reside in the substance of the rules themselves?

1. Is war a rule-governed activity at all? There is a well-established doctrine that it is not. Perhaps its most concise formulation is to be discovered in the theme developed by Michael Walzer that 'war is hell'.[22] If this perspective has meaning, it surely resides in the notion that war is an ordeal of suffering, violence, and endurance in which any act thought necessary to further the cause, either directly or indirectly, is permissible. Indeed the issue of permissibility can scarcely arise.

Such a doctrine seems unacceptable on at least two counts. As a matter of principle, the idea of war as indiscriminate violence suggests an image of violence as an end in itself which sits uncomfortably with the prevalent concept of war as an end-related activity—that is one which is directed to the attainment of some objective. If war is goal-oriented, it would seem reasonable to deduce from that fact at least that the means of war do not contradict the purposes for which the war is being fought. This is not itself a major restriction on the conduct of war. If the purpose of war is to attain victory, then the concept of war would be highly permissive of the manner in which it is to be conducted. None the less, even if not restrictive in content, the rudimentary idea that war is to be directed to the purpose of victory introduces the thin end of an important principle. It takes us at least one small step away from the 'war is hell' doctrine and this one small step is sufficient to distance us from the absolute denial that any rules apply in war. Even a strictly instrumentalist view that all necessary means to the achievement of victory are permissible is itself implicitly restrictive inasmuch as it limits hostilities to measures that are deemed to be 'necessary' and this implies that measures of violence superfluous to the attainment of military objectives, and hence not strictly necessary, may themselves be proscribed. Instrumentality, by itself, does little to restrict warfare but it offers a foundation upon which a more elaborate concept of war, as a political or moral instrumentality, can be constructed.

[22] *Just and Unjust Wars* (Basic Books, 1977).

The doctrine of unrestricted violence seems also to be untenable on more pragmatic grounds. The history of philosophical speculation about war in a range of diverse cultures, as well as the history of the practice of war, seems to be based on the contrary assumption. This is not to say that in reality war has always, or even mostly, been conducted with restraint. What can be said, however, is that there is a sufficient body of practice which recognizes principles of restraint, even while implementing them poorly, if at all, to suggest that there is a universal aspiration to impose restraints on the conduct of war. Indeed, the military codes of all contemporary national armed forces enshrine the doctrine that there are rules of war and that these rules are to be observed by the members of the armed forces in the course of their military duties. Such national codes embody disciplinary measures for those military agents who transgress the bounds of accepted military conduct. If war were truly hell, it is difficult to imagine why such codes would be required because even the principle of military efficiency, in terms of which much military discipline is justified, would have little value.

Moreover, the 'war is hell' doctrine is frequently couched in the language of the right of the state to resort to such means as are necessary to safeguard its own interests, or its very existence. But this appeal has its own inescapable corollary. 'The right of a state to make war', Nardin insists, 'presupposes the existence of rules from which that right can be derived' which, in turn, are a part of the 'authoritative common practices' of international society.[23]

2. If it is rule-governed, what is the nature of the rules which govern it? It is not sufficient to say that war is a rule-governed activity. We need also to know something of the nature of the rules which are to apply. Once again, we are led to the conclusion that different concepts of war imply different rules for its conduct and, even where the substance of these rules may partially coincide, what sets them apart is the motivation which leads to their adoption. Of the possible types of rules, we can distinguish five:

(a) *Rules of military efficiency* If the predominant concern with war is how to win it, or possibly how best to avoid losing it, then it might be said that from the concept of war we can deduce only certain principles of military necessity. What determines whether any act of war is to be undertaken or not is the purely technical consideration of its military

---

[23] T. Nardin, *Law, Morality and the Relations of States* (Princeton University Press, 1983), 279.

contribution to the course of the war. In these terms the rules of war are to be judged by the military professional. In accordance with such rules looting may be prohibited, not because it is an infringement of some extrinsic ethical code, but simply because it gets in the way of the efficient conduct of the war and is bad for the discipline on which an efficient army depends. Various codes of chivalry, and of professional solidarity, may ultimately be derivative from this requirement.

(b) *Rules of political instrumentality and proportionality* If it be true that war is the continuation of the business of politics by other means, then war is not an autonomous activity but one constrained by the requirements of policy. This demands at least that the conduct of the war be directed to the political objectives of the war and also that the means employed be in some proportion to the expected political gains. In accordance with rules of this nature, the goals of war would prevent us from devastating a territory which we sought to annex and would require us not to expend political resources in excess of the fruits of our military labours. Whether proportionality means more than 'economy of force' is an issue that is to be found at the meeting place of the just and limited war traditions which are reviewed below.

(c) *Rules of utility* In the prosecution of war there may be a common advantage to both sides in removing certain inconveniences and burdens. For instance, it can be reasonably anticipated by both sides that they will incur casualties in military operations and that some of each side's forces will be taken prisoner. To the extent that this is so, there may be no unilateral advantage in denying assistance to the wounded, or in killing or mistreating those taken prisoner. On the contrary, there may be mutual advantage, in terms of enhancing the morale of the respective armed forces, in being able to assure one's own combatants of humanitarian treatment if such unavoidable circumstances should befall.

(d) *Rules of positive law* It may be objected that most of these other kinds of rules are either destructive of any enforceable restraint in war or make appeal to principles which are themselves uncertain in foundation and doubtful in applicability. Accordingly, as with other aspects of inter-state relations, it might be contended that the most appropriate manner of subjecting war to observable rules for conduct is by enshrining it in a code of positive international law. The law of war, and the law of armed conflict more generally, has arisen in response to the need to codify and consolidate the rights and duties of belligerents.

(e) *Rules of morality* The final set of rules assumes that war, far from being hell, is a human activity, subject as all human activities to moral norms. Whereas the 'war is hell' doctrine implies that the last moral choice is that of going to war, and that morality has nothing to say to the realm of war, this concept of war is firmly grounded in the idea that war itself is a moral condition. Simply because they are in a state of war, belligerents do not cease to be human beings and they continue to be related to each other by the same moral rights and duties which prevail in other areas of human intercourse. Moreover, to the extent that there is assumed to be a moral purpose in war itself—that is, that it is undertaken for the realization of 'values'—it is further assumed that it would be self-contradictory to infringe moral values in their pursuit. In other words, from the very concept of war as a moral instrumentality, we can deduce certain principles for the conduct of war as it would be inconsistent to have the means of war subvert war's moral purpose.

3. Does the value in having rules of war reside in the fact that there are rules, or does it reside in the substance of the rules themselves? Beyond the issue of the specific nature of the rules and of the concept of war from which they are derived, there is a more general issue which will recur in a variety of contexts. In any discussion of the philosophy of war and of the rules which may be thought to apply in the conduct of it, we are confronted by a choice whether the value of the rules of war derives from their very existence (their precise content being of lesser, or possibly no, importance) or whether the value lies in the specific content of the rules (having no rules would, it is deemed, be preferable to having 'bad' rules).

For purposes of illustrating this choice, we might present a formalized contrast between a crudely utilitarian view of the rules of war and a moral theory of those rules. According to the first conception, any rule which diminishes the inconveniences of war and which both belligerents can be prevailed upon to observe, is worthy of respect. Even if there is no particularly compelling moral basis to the rule, we should welcome it if its consequence is to reduce, however arbitrarily, the destruction and suffering of war. A possible *reductio ad absurdum* of this position would be that it is impermissible in war to kill anyone with blonde hair. It would be difficult to articulate the general moral principle from which such a rule derives and, to that extent, it could be said that the substance of the rule is itself morally neutral. Nevertheless, on utilitarian grounds, it might be deemed better at least to save blondes rather than to save no one at all. In consequence, there might be value in respecting certain rules of war which themselves have no moral content.

Such an example is fanciful. Its practical correlate in real war is the status of non-combatant immunity. Is this an expedient attempt to protect from the inconveniences of war a morally arbitrary group of people? Or does the principle embody intrinsic moral value? A case can be made that the principle of non-combatant immunity represents a false attempt to build a contingent practice of war into the very concept of war itself by making deliberate attacks upon civilians unacceptable by definition. Historically the clash of armies has preceded the onslaught on civilians. Now, in the age of air power, such a preliminary military clash is no longer required. Is non-combatant immunity, therefore, no more than an outmoded technique of warfare, presented under the guise of the very essence of warfare itself?

The contrary position would be that the value of the rules of war must be found in the specific substance of the rules themselves. What lends credence to a theory of non-combatant immunity, is that it is possible to develop a general moral argument in terms of which respecting the rights of non-combatants makes moral sense, in a way in which respecting the rights of blondes could never do.

What may seem to be a very abstract point soon reveals itself to be a matter of supreme practical importance. For instance, in seeking to develop rules for the conduct of war, be it for military, political, utilitarian, legal, or moral reasons, we are necessarily confronted by the problem of translating principle into practice. At this point a choice often becomes inescapable: for instance, should we aim to establish rules that are theoretically consistent and coherent in terms of content, or, rather, should we aim to satisfice, by resorting only to principles that are observable in practice. What is the value of a rule of war which, with the best will in the world, cannot be translated to the battlefield? Is it not a higher moral wisdom to accept distinctions which, even if arbitrary in themselves, have the virtue of being realizable?

The dilemma can be demonstrated in the following examples. It might be thought desirable that our weapons be turned against those only who bear subjective moral guilt for the conditions which have brought the war about but, however commendable in theory, the defect of such a position is its manifest impracticability. How is this distinction to be recognized in practice in the heat of battle? By way of contrast there are at least some observable external characteristics, however ambiguous, in terms of which non-combatancy might be judged. Admittedly, some non-combatants might be more directly responsible for the war than some 'innocent' members of the armed forces. None the less, the greater moral force of the former principle is counterbalanced by the greater claim to observability in practice of the latter.

This is but one way of saying that the saliency of a distinction may be more important than its moral basis. This point is perhaps best illustrated in the context of recent discussions of the 'uniqueness' or otherwise of nuclear weapons. We should be clear that in such discussions there are two issues at stake which ought properly to be distinguished. The first is whether there is indeed any basis of distinction between nuclear weapons and other kinds of weapons in terms of their intrinsic nature and the kinds of physical effects that they have. Accordingly, some might discriminate against nuclear weapons by saying that they are intrinsically immoral or politically unusable. To say either of these things, of course, does not by itself resolve the second issue which is whether or not, regardless of the intrinsic properties and qualities of these weapons, the distinction between them is useful solely for the reason that it is salient and readily detectable. Even if using nuclear weapons is no more or less objectionable than using other kinds of weapons, the nuclear threshold has a value simply because it exists and can be recognized. This fact is more important for the conduct of war than the moral or political coherence of the principle upon which such a distinction might be based.

What this introductory chapter has done is to establish three areas of analysis. The first is the contention that there is an intimate connection between the very concept of war and the kinds of rules which might be set down for its conduct. The second is that the rules of war themselves vary in motivation depending upon the specific concept of war from which they are derived and it is possible for any one rule of war, such as sparing the lives of prisoners of war, to be explained in many ways depending upon a variety of military, political, utilitarian, legal, and moral assumptions. Finally, over and above the issue of distinguishing the nature of the dialogue which informs the specific content of the rules of war, there is the further complicating factor that the rules of war may have value beyond the substance of the rules themselves. In elaborating a philosophy of war it is of major importance to lay bare our scale of priorities and to demonstrate how these priorities are to be related to each other.

There is no single system of ideas for doing so. Our intellectual heritage on the subject of war and the manner of its waging derives from two principal traditions which will be reviewed in the succeeding chapters. These are the just war and the limited war doctrines. Each establishes differing rules for the conduct of war but, given the extensive overlap between them, they are more interesting for the differing motivations which underlie these rules than for the substance of the rules themselves.

These traditions offer a convenient point of entry into the analysis of the concept of war and the relationship between concepts and the general principles for the conduct of war that can be deduced from them. They thus focus on our central initial problem which is that of an adequate philosophical orientation towards the nature of war itself. This is a problem which is especially acute at the present time, given the destructive potential of contemporary weaponry and the prospective aberrations from the conceptual norm which this might imply. As a recent collection of essays on the evolution of strategic thought has suggested, the central unease of the present time penetrates to the conceptual core of war: 'Many people have reacted to the destructive power of nuclear weapons by rejecting the concept of war in general, and consequently feel that the nature of war itself no longer requires investigation.'[24]

Paret's diagnosis is tellingly correct but the flight from reality which it addresses is finally a counsel of despair and of evasion. It is precisely because of the incongruities between our traditional concepts of war (although the 'traditional concept' is itself of relatively recent origin) and the horrendous military means now available that the philosophical investigation of war must be carried forward. To abandon the task at this point would contribute nothing to our mortal salvation and serve only to hasten our intellectual bankruptcy in the meantime.

[24] P. Paret (ed.), *Makers of Modern Strategy* (Princeton University Press, 1986), 7.

DOCTRINES OF JUST WAR

IT is not possible to speak of a single doctrine of just war; nor can we point to the lineal development of a single idea; nor can we talk of the doctrine having a continuous history. At best, just war doctrine is a set of recurrent issues and themes in the discussion of warfare and reflects a general philosophical orientation towards the subject. It is, in the words of one recent exposition, 'a mosaic of thought fashioned by theologians, philosophers, jurists, statesmen, and soldiers'.[1] But to say this little is to say rather a lot. The reach of just war discussion is extensive and it is difficult to make any intelligent comment upon the nature of war, and how to wage it, without recourse to ideas and language derived from the just war tradition. Moreover, it is a living tradition that has demonstrated considerable persistence and adaptability and should be regarded as 'a practical body of moral guidelines applicable to real life, not a museum piece to be preserved for its own sake'.[2]

The significance of just war doctrine for the present discussion is that it offers a compelling instance of the intimacy of the relationship between the concept of war and its permissible means of conduct. By starting from the premiss that war is sanctioned for specific purposes alone, just war doctrine entails necessary restrictions upon its mode of prosecution. 'The principle forbidding indiscriminate warfare', Ramsey holds, 'pertains to the nature of warfare itself and its own proper laws—so long as this human action remains, by the skin of its teeth, a rational activity at all.'[3]

Immediately, however, questions about the continuing relevance of the doctrine force us to contemplate the nature of the concept of war itself. Can a set of principles which was established and elaborated in medieval Europe, and which came to be transmuted into the early theories of international law, claim contemporary relevance in the age

---

[1] D. L. Davidson, *Nuclear Weapons and the American Churches* (Westview Press; 1983), 13.
[2] W. V. O'Brien, *The Conduct of Just and Limited War* (Praeger, 1981), 5.
[3] P. Ramsey, *The Just War* (Charles Scribner's Sons, 1968), 164.

of nuclear weapons? If it is true, as many strategic analysts have claimed, that the concept of war has undergone a 'Copernican' revolution with the introduction of the new means for its conduct, does this not at a stroke render outmoded the application of just war considerations?

The full complexity of the interrelationship between the concept of war, the ends for which it is fought, and the instruments by means of which it is pursued begins to manifest itself at this point. The case has thus far been advanced that ideas about the proper conduct of war are intimately, and necessarily, associated with the very concept of war itself. Unhappily, there is also a circularity to this proposition because the concept of war, in part at least, derives from changing instruments for its prosecution—otherwise the nuclear revolution in thought about war could not have occurred. The debate about the contemporary relevance of just war doctrine consequently proceeds at two levels. At the first, and more superficial, the question is whether principles of restraint derived from a technologically primitive environment are of continuing utility: at the second, and more profound level, the issue at stake is not simply that of controlling the instruments of war but of deciding whether our concept of war is recognizably the same as that which developed in the earlier period.

Origins

Linguistically, the term *bellum iustum* derives from Roman law and is itself an interesting blend of a religious and legal idea. Strictly, the Roman notion of a *bellum iustum* represented a war which was initiated and executed in adherence to the necessary formalities. Already then, with this idea, there is in being a close association between the idea of war and war as an activity defined by its own rules. But the function of the rules of the Roman just war went beyond mere legal niceties. The reason why it was necessary to adhere to the formalities was that only a war which so conformed would be blessed by the gods and by good fortune. Legal form was therefore an adjunct of religious practice: only that war which respected the requisite forms would be just. As Russell describes the practice of the Republic, the issuing of a *repetitio rerum* was a demand for satisfaction which, if unmet by the foreign power, entitled the fetial priests to make a formal declaration of war. 'By this procedure', Russell concludes, 'the just war had a religious as well as a formal aspect, for by adhering to the *ius fetiale* the Romans hoped the gods would aid them in battle.'[4] It might be said, therefore, that in this Roman conception it is scarcely possible to make a separation between the rules of engagement, viewed as legal principles, and the ethico-religious function of war within a wider value system: not to respect the formal rules of the activity of war would be counterproductive in that it would negate the purpose for which war was undertaken and risk alienation of the gods. To this extent, the practice of Roman just war derives immediately from an understanding of its religious connotations.

However, the origins of just war doctrine, as we have come to know it, are normally traced from the early beginnings of the Christian Church and from its conversion away from a posture of pacifism. While the early Church had been strongly pacifist, its adaptation to post-Constantine 'establishment' brought with it a worldly acceptance of the need to defend the spiritual realm within the temporal and it is in this acceptance that just war origins are to be discovered.

The form in which the question about resort to war was phrased was itself revealing. What concerned the early Christian writers, from Augustine onwards, were the conditions under which a Christian might justifiably resort to arms. The assumption underlying this question was that resort to arms is not normally legitimate and that it requires exceptional circumstances before the normal prohibitions are lifted. These exceptional circumstances would be present, for Augustine, when a moral wrong had already been committed and, thereupon, it would be the duty of the Christian to take martial action, even with a heavy heart. His reasoning was simply that 'war and conquest are a sad necessity in the eyes of men of principle, yet it would be still more unfortunate if wrongdoers should dominate just men'. To Augustine 'just war is action designed above all else to restore a violated moral order'.[5] In other words, the resort to force is an exceptional condition requiring justification and the onus of establishing just cause rests firmly upon the person who would resort to it. The tendency of the doctrine was, therefore, not to expel war into the moral void but rather the contrary. The seemingly immoral class of killing and destruction was to be related to the moral universe by insisting that it could be rendered legitimate if certain specified conditions were met; if not, then not.

It is immediately apparent that the requirement of justification is

[4] F. Russell, *The Just War in the Middle Ages* (Cambridge University Press, 1975), 6.
[5] R. S. Hartigan, 'Saint Augustine on War and Killing', *Journal of the History of Ideas*, Apr./June 1966, 199.

permissive as far as is necessary for the accomplishment of the aims of the just resort to arms but that beyond this point all remaining violence continues to be proscribed. This means that the philosophical point of departure for just war theory is that war is an activity circumscribed by certain rules and that when the just demands of war have been met, the licence for any additional violence is revoked. The very concept of just war therefore demands an appropriate means for its conduct: justification means continued prohibition of that which is not justified. Accordingly, as James Turner Johnson has suggested, just war *is* limited war: 'That just war theory permits Christians to participate in one particular form of violence under certain specified conditions is clearly true; yet such permission goes hand in hand with limitation.'[6] Ramsey argued similarly:

> The justification of participation in conflict at the same time severely limited war's conduct. What justified also limited! Since it was for the sake of the innocent of the earth that the Christian first thought himself obliged to make war . . . he could never proceed to kill equally innocent people as a means of getting at the enemy's forces.[7]

These observations are evidently correct but precisely how closely the doctrines of just and limited war are to each other will be reviewed at a later stage.

Content

Discussion of just war doctrine normally revolves around two principles which are variously related. The first is the *ius ad bellum* which specifies the conditions under which war might justly be undertaken at all. It is the dialogue about those causes which would justify the resort to arms in the first place. The second principle is the *ius in bello* which specifies how the war is to be justly waged when once it has been embarked upon. It is, therefore, the debate about the just means of war and is concerned with such matters as the proper targets of warfare and the instruments that are chosen to attack them.

In terms of substance, the *ius ad bellum* has been variously thought to specify a number of conditions to be met in undertaking a just war. These have included some rudimentary idea that war is not to be

---

[6] J. T. Johnson, *Just War Tradition and the Restraint of War* (Princeton University Press, 1981), pp. xxvi–xxvii.

[7] Ramsey, *The Just War*, p. 143.

undertaken lightly (reinforcing the conception of it as a situation requiring permission) and should therefore be entered into only as a last resort when other measures to resolve the conflict have been expended. Secondly, and crucially, just war theorists have insisted upon the declaration of war being made only by a legitimate authority, again emphasizing the complex relationship between moral precept and political culture. Since this authority was to be the sovereign prince, this had the effect of regulating only wars between those princes.[8] Thirdly, war should be resorted to only for a just cause, which has been imprecisely defined but which is usually thought to focus upon requirements of self-defence, restitution, or retribution. Latterly, extended conceptions of just cause, including a right to pre-emption and to humanitarian intervention, have been substantially constrained by the just war heritage in contemporary international law. Finally, some theorists have insisted upon a reasonable prospect of success as a condition of a just war. Without such a prospect, the evils of war would simply be added to the evils of the unjust peace.

Summarily, the principle of *ius in bello* dwells upon the ideas of proportionality and discrimination. Both set bounds to the level and nature of force that might be employed in war, by stipulating that the war should not contribute more harm than good and that, in its prosecution, the targets of war should be selected in accordance with an intelligible moral principle, predominantly but not exclusively that of combatancy.

Each of these principles is a matter of considerable philosophic interest in its own right. The search for overarching principles which would justify the resort to war has been both a search to add moral weight to the enterprise of war and also a search for a restrictive principle which would outlaw those wars that did not meet the criteria of justness. The search for overarching principles which would justify the means of war has likewise served both to remove guilt from the soldier engaged in a just enterprise and also to restrict such means as have been considered unjust. Individually, therefore, the twin elements of just war doctrine have demonstrated the dual function of just war theory both as a form of permission and as a form of restraint.

While the principles have intrinsic interest apart, it remains none the less true that by far the most interesting issues associated with just war doctrine have emerged from attempts to see how the two are related to

---

[8] Johnson, *Just War Tradition*, p. 61.

each other. In Walzer's words 'the dualism of *ius ad bellum* and *ius in bello* is at the heart of all that is most problematic in the moral reality of war.'[9]

There is both harmony and conflict in the relationship. Indeed, like the present-day superpowers, the two principles might be said to enjoy an 'adversary partnership'. On the one hand, the relationship is a harmonious partnership inasmuch as one would expect continuity between the ends of a war fought for certain values and the means of war; a war *for* certain values should express these values in its own conduct. Accordingly, it would be a perverse state of affairs if a war undertaken for the sanctity of human life were to express itself in practice by flagrant disregard for this principle in excess of that strictly necessary to accomplish the war's goals.

Beyond this, the adversarial element of the relationship is the more conspicuous. In the case where it is believed that there is only one just party to the conflict, that is, one party whose cause is just, why should that party be restrained in its prosecution of the war in the same manner as the unjust party? Since war is not a game, and we are not indifferent to its outcome in devising the rules which govern it, why should we prejudice the result by expecting the party which is fighting for a just cause to fight in such a way that it may lose? If the cause be just, is it not immoral to run the risk of losing by being overly scrupulous about the conduct of the war? Such is the force of this antagonism that a leading scholar of the laws of war has contended that, historically, preoccupation with just cause of war militated against the development of laws for its conduct: 'And so the importance attributed to the idea of just war throughout the Middle Ages and well into the seventeenth century undoubtedly delayed the appearance of any body of rules restraining the more barbarous practices of warfare.'[10]

The point is no sooner made than the obverse relationship suggests itself. Is it not possible that the right cause of a belligerent can be squandered in its unjust conduct of a war? Can we make any final determination on justice simply on the evidence of the cause which has precipitated hostilities or do we not need to await the course of the fray to see which party acquits itself the better in the manner of its waging of the war? Is there not the danger that a doctrine of absolute right for the just party will become corrupted into a doctrine of absolute licence

⁹ M. Walzer, *Just and Unjust Wars* (Basic Books, 1977), 21.
¹⁰ G. Draper, 'The Idea of the Just War', *Listener*, 14 Aug. 1958, 222.

which, in turn, will pervert the basis of just war doctrine itself? Melzer draws attention to this view in recalling of certain acts of war that 'not only are they criminal in themselves but they also render criminal the war as a whole'.[11]

Development

The elaboration of just war doctrine at the hands of Augustine of Hippo set the framework of discussion for the ensuing millennium. What was unmistakable about Augustine's treatment, an aspect that was to be characteristic of most just war theorizing throughout the medieval period, was its preoccupation with the *ius ad bellum* to the almost total exclusion of any explicit consideration of the conduct of war itself.

In Augustine's treatment, the permissive element in just war theory is much more conspicuous than the restrictive element. Dominated by the notion that there was a divine duty to punish the evil-doer, Augustine carried his concept of war over into a tendency to make its prosecution limitless. Since there is a duty to punish wickedness, and since war is a chosen instrument for doing so, there can be no limit to such duty nor any corresponding right to escape divine retribution.

Part of the reason why, during most of the medieval period, the *ius in bello* struggled vainly to achieve an independent existence was the lack of precision in the purpose of war itself. Because the concept of war lacked clear definition, medieval scholars and canon lawyers were ill-equipped to distinguish clearly a body of theory about the proper means for its conduct. The conflation of a range of medieval ideas is reflected in the extent to which a number of just causes—defence against wrongful attack, avenging injuries wrongfully received, and divine punishment of wickedness—are all run together as if interchangeable. Russell notes that the contemporary formula of *ulciscuntur iniurias* contained ambiguities 'for it could entail merciless and unrestrained revenge for a trivial injury, restrained defence against hostile attack, recovery of stolen goods, or even the punishment of evil-doers'.[12] However, it should be apparent that each of these purposes of war would differ in its degree of permissiveness in relation to the war's conduct.

Even as late as the writings of Aquinas, the theme of the *ius in bello* is

[11] Y. Melzer, *Concepts of Just War* (Sijthoff, 1975), 95.
[12] Russell, *The Just War*, pp. 66–7.

implicit at best. What Aquinas was to succeed in doing was to pull together various strands of canonical writing on the subject of war and to present an authoritative account of the sundry criteria which constituted the just cause of war. According to Aquinas's account, there were three essential conditions which had to be met before a war might be considered just. These were that the war be fought with proper authority, that it have a just cause, and that it be prosecuted with a right intention.

While consideration for the means of waging war was seldom explicit in medieval treatment of the just war, it is by no means altogether absent. A clear association can, for instance, be drawn between Aquinas's requirement for right intention and a principle of just conduct. It was undoubtedly an element of right intention that the object of war was to restore a better peace and the just warrior would, accordingly, have to the front of his mind the realization that this was his purpose.[13] Adherence to such a principle would preclude him from such measures in war as would serve to undermine the restoration of a just peace and, to this extent, it is clear that the criterion of right intention is to be understood as a limitation upon the conduct of war even though it is framed as part of the *ius ad bellum* and not as a separate principle for the conduct of war. It might be argued, indeed, that the association was so clearly assumed as not to need to be explicitly stated. Since war was a limited activity, and since what was justified was only that which was strictly necessary to its purpose, there was no felt need to proceed to elaborate a separate set of principles for its conduct. The umbrella concept of right intention was regarded as sufficient to remind practitioners that what could be legitimately done in war derived from the purpose of war itself and, provided war was entered into with this right intention of restoring a just peace, the conduct of war would properly take care of itself. 'The concept of right intention', Johnson reminds us, 'can properly be conceived as an important source for the *jus in bello* in Christian theological thought.'[14]

It should not be thought that the influential body of medieval just war doctrine was purely ecclesiastical in origin. Rather, it was a confluence of various streams—religious, chivalric, and secular. The chivalric input reflected the knightly code of a warrior guild which demanded that the 'professional' warrior conduct himself in certain

---

[13] J. D. Tooke, *The Just War in Aquinas and Grotius* (SPCK, 1965), 23.

[14] J. T. Johnson, *Ideology, Reason and the Limitation of War* (Princeton University Press, 1975), 41.

ways and confine his war-making to other members of the guild. It was also a reflection of the code of honour which required the knight, as a member of his knightly class, to display the virtues appropriate to his station. There was, of course, a self-serving dimension to the practice of honour, as the code of confining hostilities to other knights was but another way of maintaining the exclusive right of the knight to be a bearer of arms: honourable conduct in war was but one way of maintaining the guild.

If this was true of the chivalric input into just war doctrine, it is the more readily recognizable in the case of the relationship between that doctrine and the secular authorities. What the knightly code of combat did for the exclusive warrior caste, the just war requirement of proper authority did for princely authority: it defined the exclusive right of the prince to enter into the state of war. It had thus become part of the concept of war that it be waged only between sovereign princes and that there could be no other claimant to public violence. All other resorts to force were deemed to be merely private in nature.

We can see in these various relationships the mutual dependence and servicing which existed between the sundry feudal corporate structures: just war doctrine was that fusion of religious, chivalric, and secular ideas which welded the components into an integral whole. It gave spiritual sanction to the military enterprise, rendered knightly protection to the Church and its property, and reinforced the political authority of the prince as the only legitimate source of military action. This said, it is clear that just war doctrine had become much more than a set of philosophical precepts for the conduct of war but was, indeed, a fundamental part of the medieval social fabric itself.

It is for this reason that some critics have dismissed the ethical value of the medieval just war heritage. Rather than serving as a model for our own ethical appreciation of war, it has been rejected as being no more than a system of corporate self-interest aimed at the maintenance of a particular social status quo. As Phillips has noted:

Thus the historical context which saw the full bloom of *bellum justum* was one of almost perfect accord between prudence and morality, at least prudence as perceived by the ruling order. And there, of course, lies the problem. One of the most serious criticisms to be raised against *bellum justum* is that it is precisely an instrument devised to protect some favored form of warfare from encroachment by the unorthodox.[15]

---

[15] R. L. Phillips, *War and Justice* (University of Oklahoma Press, 1984), 11.

During the seventeenth and eighteenth centuries, the mainstream tradition of just war theorizing was to be increasingly secularized as it became, in practice, part of the language of the emerging body of international law. Indeed, during this period, the major developments in the history of just war were wrought at the hands of the international lawyers who became the main repository of the tradition. At the outset, the corpus of international law derived from shared assumptions about natural and divine law and spoke a language common to the just war tradition. The momentous changes of this period, however, witnessed the gradual transformation of the content of that law and, by association, of just war ideas as well. The style of international law writing during these two centuries reveals the impact of the disintegration of the medieval ideal of a unified Christendom, of the schismatic forces of Reformation and Counter-Reformation, and the transition from a spiritually unified concept of Europe to one that is secular and divided into territorial units marked by the new doctrine of state sovereignty.

Paradoxically, it was the collapse of the medieval order which propelled the search for new sources of restraint in the conduct of warfare. In the face of the emerging and powerful military structures of the state, some code of moderation was required to take the place of the disintegrating medieval normative order. At the same time, the political fractures that were appearing made the search for such a new value system that much more difficult. What was the universal norm that could be appealed to to moderate the excesses of war in the wake of the new doctrines of sovereignty and *raison d'état*? It was the task of the international lawyers to devise a set of principles in terms of which the new sovereign states might coexist with each other and regulate their intercourse even in a state of war. The solution that they eventually offered was a movement away from the 'necessary law of nature' of the late medieval period and towards the 'voluntary law of nations' which was to dominate the practice of the nineteenth-century state system. In the process, just war ideas had to be translated into positive law.

This transition can be illustrated with reference to two aspects of just war theory. In the first case, there was to be a pronounced, if gradual, change in the substance of thinking about the question of *ius ad bellum*. Secondly, and deriving from this development, there was to be an associated shift of emphasis away from the *ius ad bellum* in the direction of increased emphasis upon the *ius in bello* which began, for the first time, to take on a substantial life of its own.

The change in the substantive content of the *ius ad bellum* can be demonstrated by looking at the question of determining which is the just party in a war. Initially, there was some tendency in the medieval doctrine to assume that, at best, there would be only one just party. It was a contradiction for both parties to be pursuing a just cause at the same time as one, if not both, must be in error. Such a determination was easily made when the notion was that a party was just in the eyes of God. By what means, however, was such a determination to be made by hapless humans? How, in other words, with the secularization of the international law of war, could a standard for discerning justice be devised that was not dependent upon assumptions of an all-seeing, all-knowing deity? The answer was to be a progressive slide away from the initial position, resulting finally in the abandonment of the search altogether.

The doctrine was partially diluted by the attempt to distinguish between a subjective and an objective concept of justice. While objectively only one side could be just, subjectively both parties might believe they had a just cause. Since mistaken belief was a lesser fault, it was considered equitable that the party which was objectively unjust, but which believed itself to have just cause, should still enjoy certain rights in war. It was but a small step from this position to the doctrine of simultaneous ostensible justice whereby both parties had the appearance of being in the right.

If this wrestling with a new method of determining justice in war was one response to the secularization of just war doctrine, the other was the attempt to divorce the question of justice from the legal effects which were consequent upon the state of war itself. That is to say that the legal code governing the conduct of hostilities was to be viewed quite independently from the question of which side had the greater justice in entering upon the state of war in the first place. The emphasis was upon equality of legal treatment, regardless of degrees of justice in cause. This was little more than a pragmatic response to the perceived difficulty in providing a universal secular standard in terms of which justice of cause might be measured. Thus the easier route was to separate the two issues from each other and to recognize the sovereign rights of the state in war as an issue apart from the reasons which had precipitated the hostilities.

The ultimate conclusion of this tendency, which was approached in the eighteenth century and finally realized in the nineteenth, was that international law denied itself any authority for pronouncing upon the substantive issue of justice of cause. It was no longer deemed to be the business of international law so to pronounce because this was but a fragmentary remnant from a natural law tradition that was no longer credible in the age of legal positivism. Since there was no independent and universal standard by which justice of cause might be determined, international law simply had to recognize the reality that the sovereign state was the only authority which could pronounce upon its own cause for going to war. As the eighteenth-century jurist Christian Wolff was to argue 'the question of justice of wars falls outside the pale of positive law'.

By the nineteenth century, the thrust of just war doctrines, as embodied in international legal thinking, was in the direction of a return to the more formal Roman law notion of *bellum iustum* from which it had originally derived. That is to say that the provenance of international law lay in determining whether the correct legal forms had been observed and that the parties engaged in war had the legal capacity to do so. Provided the state had a right to make war, which the sovereign state clearly had, the substantive issue of the relative merits of the cause did not arise as a matter of law, but simply as a matter of international politics.

The same development can be traced by stating that there was a progressive shift away from *ius ad bellum* and towards *ius in bello*. This was based on the recognition of the realities of the political situation inasmuch as to emphasize just cause, without any hope of securing the agreement of the rival states, would simply interfere with the business of seeking to regulate, and hence mitigate, the practice of war itself. Hence, if war was to become a practice regulated by positive law, it was necessary to forswear involvement in the underlying causes of it. Thus Vattel could write in the mid-eighteenth century that 'the first rule of that law . . . is that regular war, as regards its effects, must be accounted just on both sides. This principle . . . is absolutely necessary if any law or order is to be introduced into . . . war.' Failing this, the continued preoccupation with questions of *ius ad bellum* would prevent any progress being made in the more promising area of *ius in bello* restrictions. Failure to regulate the resort to war should not also mean a failure to regulate the conduct of war. In any case, there was some intuitive ground for believing that the latter might be the simpler task for just war theory. Ramsey suggests that '[i]t may well be the case that natural reason falters in attempting to make large comparison of the justice inherent in great regimes in conflict but is quite competent to deliver verdict upon a specific action that is proposed in warfare.'[16] It was for this reason, because the international lawyers sought to minimize the impact of war, rather than seeking to abolish it altogether, that Kant was to accuse them of being 'miserable comforters'.

Unlike the medieval scholastics, therefore, the international lawyers dealt explicitly with the proper conduct of war itself. In this respect, their efforts were concentrated in two major areas. First, by a doctrine of proportionality, in ensuring that the means of war conformed in scale and effect to the objectives being sought. Secondly, by a principle of discrimination, to contend that war was intrinsically a limited activity that could properly be directed against legitimate targets only.

The content of these principles, and some of the problems associated with them, will be considered at a later stage. For the moment it is sufficient to note that, in terms of the principle of proportionality, the concern was that the means of war should not be excessive relative to the specific military objectives being sought and, more generally, that the total costs of the war, in moral as well as in physical terms, should not exceed the benefits derived therefrom. In the words of Victoria 'greater evils do not arise out of war than the war would avert'. If the war failed to conform to this criterion, it could scarcely be judged just in the grand reckoning.

As regards discrimination, the theorists attempted also to construct plausible principles in terms of which the reach of battle might be limited. They tackled the problem from various angles, suggesting that immunities adhered to those not offering, or incapable of offering, armed resistance; to those whose social function was not connected directly with the war effort; and also to such intractable, and virtually inoperable, categories as those of guilt and innocence in bearing responsibility for the war. However, these immunities were seldom presented as absolute requirements for the conduct of just war. Allowance was made for the play of military necessity by such arguments as that of double effect which permitted a degree of 'collateral damage' provided that the killing of the innocent was not intended, that it was an inescapable secondary effect of a justified military action, and that the good derived from the primary intent

[16] P. Ramsey, *War and the Christian Conscience* (Duke University Press, 1961), 33.

outweighed the evil brought about by the undesired, and unintended, secondary effect. For instance, some theorists considered the execution of a siege against a town to be a legitimate military operation and allowed that some loss of innocent civilian life was an unavoidable secondary consequence of the pursuit of such a justified military objective.

The international lawyers were also perplexed by the seeming paradox of war, that to respect certain restraints in the conduct of war required a degree of co-operation, explicit or tacit, with the enemy. How could this be? As the eminent legal theorist Grotius was to point out: 'If the issue at stake . . . is worthy of war, we must strive with all our strength to win.' This, at least at first sight, seemed to be the logic of war itself and it leaves no room for the idea of co-ordinating military activities with the enemy in such a way as to civilize the business of waging war. How then is this paradox to be resolved, that the purpose of war is to defeat the enemy but the just conduct of war requires a degree of co-operation between enemies?

The dilemma manifested itself on the particular issue of keeping faith with the enemy. Was it reasonable to expect that enemies, engaged in a deadly struggle with each other, should be obliged to keep faith and, if so, what could be the source of this particular obligation? The dilemma was a poignant one, as Pufendorf was to confess: 'And yet it implies a confusion for me to demand that another keep faith with me, and at the same time avow that I intend to remain his enemy.'

The solutions offered to this particular dilemma were various. Three alone will be noted. The first was the straightforward utilitarian argument that since any moderation in warfare was itself dependent upon the assumption of trust and good faith, not to maintain good faith with the enemy would simply lead to a more cruel style of warfare than could otherwise be achieved. It therefore made sound prudential sense to give the enemy confidence that he could moderate his own operations in the reasonable expectation that his protaganist would reciprocate.

Secondly, there is the argument that war does not entail the cessation of all moral practices and therefore, as the keeping of faith is a general ethical imperative, it continues to have force unless there is a compelling reason to the contrary. At this point, the moral and the prudential arguments join forces. Since abuse of the enemy's trust may bring short-term advantage, but only at the cost of long-term penalties (since the enemy will engage in reprisals) the case for defecting from the understood and accepted conventions of warfare for reasons of military necessity can seldom be compelling. Consequently, not having been overridden, normal codes of ethical conduct, such as the keeping of promises, continue in force.

Thirdly, an argument for keeping faith with the enemy during war can be derived from the intrinsic nature of war itself. While it may be held, with Grotius, that the purpose of war is victory, and that therefore all is allowable which contributes to the best realization of that end, victory itself is but a means to an end and not an end in itself. According to the right intention of the just war theorists, the object of war is to restore a just condition of peace. It follows, therefore, that since the termination of war by any kind of agreement, such as an instrument of surrender, is itself dependent upon the operation of trust between the belligerents—otherwise how can the victor have confidence that the vanquished will not use the pause simply to restore his military forces for a renewed onslaught?—the keeping of faith in war is a necessary means to the restoration of peace. And so it was argued that the concept of war, in so far as it was predicated on the goal of achieving a just peace, was not contradicted by the observance and keeping of faith with the enemy during the war itself: on the contrary, the concept of just war itself demanded that such faith be kept.

Just war doctrine, often implicitly rather than explicitly, has also had to face the issue of the applicability of restraints in war to all forms of belligerence. Do just war restrictions apply to all enemies, or only to those who share certain cultural values? It is a doctrine for a homogeneous international society or can it operate where heterogeneity is the dominant trait? It has to be said that, in practice, the distinction was often made between wars amongst 'civilized' states and those against the 'savages'. In the latter, the demands of *ius in bello* were, where they were thought to apply at all, more loosely regarded.

Revival in the Twentieth Century

After a long period of eclipse during the later eighteenth and nineteenth centuries, the just war tradition enjoyed a conspicuous revival during the twentieth. There are a number of reasons which .account for this new development.

Perhaps the most general, but also the most significant, has been the major change that has taken place in the attitude to warfare during this century. While war had been accepted as a necessary adjunct of the

European state system during the nineteenth century, at least in part because of the low physical cost of the wars that had until then been experienced, the twentieth century witnessed wars on a dramatically larger scale which prompted questions about the continuing acceptability of war as an institution of international society.

The First World War provoked a striking reaction against the idealistic view of war. Both the duration and the intensity of the fighting stimulated a new desire, not simply to control the horrors of war once undertaken, but to go further and try to abolish war altogether from international life. Thus when the new institution of international order was created in the shape of the League of Nations, it was designed with the normative task of preventing wars—apart from those which could be justified on defensive grounds or which were required as demonstrations of the international community's resolve to implement the new code of collective security.

By seeking to distinguish between aggressive wars, which were to be outlawed, and defensive wars, which were acceptable if strictly necessary, the League was in fact restoring the substantial distinction of the *ius ad bellum* albeit in a legalistic, rather than moral or religious, form. Appearances apart, however, there can be no denying that the League presided over a new attempt by international society to implement in practice a distinction between just and unjust wars. It was an attempt to create 'the necessary international structures and procedures to give effect to a revived and revised just war idea'.[17]

The reawakening of just war ideas in the present century was also prompted by the 'totality' of the two world wars. Partly as a result of dramatic changes in military technology, but also in conjunction with the new political and ideological fervour which has developed in the course of them, the world wars witnessed the overthrow of the practical restraints on war which had operated in the previous century. The distinction between soldier and civilian was eroded in practice and there were also notable violations of the codes for treatment of prisoners of war. It was as a reaction to the ferocity of such means of waging war that renewed interest in possible ways of moderating its conduct was to emerge.

It was further stimulated by the adoption by Marxism-Leninism of the language, if not the spirit, of the just war tradition. Wars of

---

[17] I. Claude, 'Just Wars: Doctrines and Institutions', *Political Science Quarterly*, Spring 1980, 92.

liberation were to be assisted because they were intrinsically just.[18] As Best has remarked, 'this marxian flood through the landscape of the law of war . . . has brought back, and deposited with a solidarity which looks unlikely to be eroded, the concept and the terminology of "The Just War".'[19]

Finally, both in the late 1950s and again in the early 1980s, the attempt to inject just war ideas into the debate about contemporary strategy has been specifically related to the issue of nuclear weapons. Their role in the ongoing strategy of deterrence, and the possibility that any conventional war might escalate to involve also the use of nuclear weapons, has led to sporadic debate about the justice of nuclear weapons and about the acceptability of strategies which depend upon their possession and threatened use.

Whether this twentieth-century revival of the just war tradition is wholly to be welcomed is a difficult question to answer. On the one hand, given the enormous political, philosophical, and physical importance of war, it is only fit and proper that people should give it the intellectual attention which it deserves. To this extent, all attempts to think carefully and precisely about the nature of war in the human experience and about its proper role and conduct, must be warmly endorsed. Given the present means of destruction, war is something about which we have no choice but to think seriously.

But the attempt to discuss contemporary warfare in just war terms may not be an unmixed blessing. Two arguments against such a revival warrant our consideration.

The first questions the applicability of a doctrine, developed in a different cultural and technological age, to the circumstances of the present times. Is there not a danger that, by using just war terminology to describe a style of warfare to which it is radically inappropriate, we do little more than create a moral obscenity? If we discuss the conduct of a nuclear war using such traditional terms as those of proportionality and discrimination, are we using the words out of context and merely putting a moral gloss upon a type of combat in which the words can have no real meaning?

Secondly, the just war tradition was nurtured in a uniquely homogeneous cultural environment. It was a product of European Christendom and, indeed, the constraints of just warfare were not

---

[18] See C. D. Jones, 'Just Wars and Limited Wars: Restraints on the Use of Soviet Armed Forces', *World Politics*, Oct. 1975.
[19] G. Best, *Humanity in Warfare* (Weidenfeld & Nicolson, 1980), 308.

thought to apply to inter-cultural conflicts in which the participants did not share the same cultural precepts, nor the same warrior-class attributes, as did the European knights. How then can this tradition apply in an international setting which is culturally heterogeneous? Is there not a further danger that, in having everyone speak what seems to be the same language of warfare in discussing what is just and what is not, but meaning radically divergent things by these words, we have a prescription for the intensification of ideological struggle rather than for its moderation. Can particularism speak the language of universalism without substantial costs to international society?

Some Problems of Justice

The attempt to analyse the conduct of contemporary warfare by means of just war categories gives rise to a number of substantial and procedural problems. Some of these will be considered at length in Chapter 5 but, for the moment, a number of general issues must be aired.

Perhaps the most fundamental is whether the logical tendency of just war theory is to regulate the instruments of war or, alternatively, to constrain the resort to war at all. In terms of the former, the concept of war must dominate our practice of it and must determine the utility, and legitimacy of war's instruments. In terms of the latter, the conduct of war is the basic reality and it is that which must govern our abstract conceptions: in the dialectical struggle between concept and conduct, it is the material reality of conduct which will hold sway and which must lead us to revise our thinking about it.

The logic of these twin arguments can be demonstrated in turn. On the basis of traditional just war principles of proportionality and discrimination, it could be contended that nuclear weapons are unjust weapons because they are incompatible with any war fought in accord with such principles. Accordingly, the logic of the exercise is to outlaw the use of such weapons in accordance with a particular conception of war as an intrinsically restrained activity.

The alternative manner of proceeding is to read back from the illegitimacy of the weapon to the illegitimacy of resort to war in which such weapons will, or might, be used. Indeed, it might be said that the general tendency of the international law of armed force has been to head in this direction by squeezing the right to resort to war by dint of the heightened consequences of doing so in the conditions of twentieth-century warfare. The *ius in bello*, on this reckoning, becomes a further source of restriction upon the *ius ad bellum*. The extreme version of this argument, as Phillips has noted,[20] is where the possibility of use of nuclear weapons in the event of escalation might nullify the justified resort to war altogether.

At base, such arguments are predicated not simply on the unacceptability of employing certain kinds of weapons but on the more far-reaching contention that the activity performed with such weapons cannot be considered to be war at all. As a recent review of various positions on the ethics of nuclear deterrence has commented, 'none of them denies the qualitative difference between war, as it has always been known, and nuclear hostilities . . .'[21] The substitution of hostilities for war marks the conceptual leap that has taken place.

As already noted in the previous chapter, our contemporary unease about various forms of violence, including the potential use of nuclear weapons, has manifested itself in uncertainty about the very concept of war itself. Not surprisingly, the contemporary just war tradition finds itself caught up in this same intellectual task and, in response to it, has adopted two competing strategies. The first is a strategy of exclusion which denies the title of war to those types of violence that are deemed to fall outside the pale. Within this framework, just war theory continues to busy itself with the traditional agenda and to exclude from its concerns those violent pursuits that do not conceptually fit into the pre-existing intellectual patterns.

The alternative strategy is that of internalization. Confronted with new martial experiences, the concept of war expands to encompass them and this results in a period of conceptual adaptation. Like the python swallowing a pig, the irregular shape of guerrilla or nuclear hostilities stands out starkly in the body of traditional thought but is then progressively, albeit slowly, digested and becomes itself a part of the tradition. This latter is the more intellectually challenging task. It may well be the case that the new violence is such a radical departure from the old as to render grotesque their amalgamation: whether the instantaneous decimation of millions of lives has sufficient in common with the staid military encounters of the eighteenth century as to warrant describing them both as war is a moot point. But neither is it clear what intellectual purpose is served by schematic apartheid in which the more virulent forms of collective belligerence are dismissed

---

[20] Phillips, *War and Justice*, pp. x–xi.
[21] F. Winters, 'Ethics and Deterrence', *Survival*, July/Aug. 1986, 345.

as being non-war situations. This may tidy the mind but achieves little by way of changing the world. This chapter concludes, therefore, with a plea for 'constructive engagement' by means of which the tradition of just war does its best to come to terms with an unfamiliar culture of violence, not to legitimize it, but in recognition of the fact that evasion will solve none of our problems: nuclear war by any other name is just as deadly.

3

DOCTRINES OF LIMITED WAR

IF just war ideas represent one great tradition in philosophical speculation about war and the means of waging it, the other major tradition is that of limited war. As with just war, doctrines of limited war are equally diffuse and derive from a number of distinct intellectual foundations. It might be thought, on first inspection, that the two traditions are polar opposites, principles of just war seeking to restrain the exercise of force whereas the impetus of limited war doctrine comes from the felt need to make wars fightable: the former seeks to constrain and the latter to liberate. This radical dichotomy is, however, an overstatement and a fuller discussion of how precisely the two are related to each other will be offered at the end of the present chapter.

This second tradition of thinking about warfare is primarily a political tradition. War itself, on this account, has its origins in the political condition of the modern state system and particular wars are caused by the political goals of security and expansion that states set for themselves. Accordingly, since war is rooted in the political process, the means of war are but an extension of political intercourse and must be related, in purpose and in scale, to the objectives for which the war is fought. Whether such a political philosophy of war leads inevitably to a doctrine of limited war, or not, is perhaps the major philosophical question generated within this particular tradition.

On the face of it, the point of departure of the political philosophy of war is radically opposed to that of the just war tradition. The extremity of the contrast is perhaps best captured in Machiavelli's sharp dismissal of the basis of that other tradition: 'Where the very safety of the country depends upon the resolution to be taken, no considerations of justice or injustice, humanity or cruelty, nor of glory or of shame, should be allowed to prevail.' Ostensibly, the language of the political tradition stands in stark contrast to the language of just war.

Machiavelli's comment needs to be placed in its historical context. What it points to are the major changes which were being wrought on the political face of Europe in the early modern period. If just war ideas had developed on the basis of a compact amongst the corporate structures of the medieval order, then the basis of this compact was being progressively eroded. The emergence of the self-conscious state embodied in its dynastic ruler, the disintegration of moral universalism, the weakening of the view of the divine grand design, the fragmentation of the Church itself, and the development of a new style of warfare associated with professional mercenaries rather than with the feudal knight, all contributed to the watershed in political ideas that was to produce a new conception of political authority, of the unit of political loyalty, and of the place of force in the shaping and maintenance of this new political order.

It required considerable athletic agility for the doctrine of just war to retain a foothold on this new ideological structure. Just war ideas had developed in the philosophical context of the universal moral community, defined in divine or naturalistic terms. The new edifice presented a frontal impediment to the fundamental assumptions of the old. In the place of the universal moral community, it offered instead a moral community which was coextensive with the individual state and between which there was nought but a moral void. Indeed the state itself, as the initiator of political order from the pre-existing state of nature, was also the initiator of the moral order itself. Between any two such self-contained communities, no moral precepts could hold. Thus the Hobbesian description of the original state of nature came to be accepted as an accurate analysis of the secondary state of nature that had now been created between the individual states themselves. If it was true, as Hobbes suggested, that '[t]o this war of everyman against everyman, this also is consequent, that nothing can be unjust', then it followed also that in the relations between states there could be nothing unjust, least of all when they were engaged in mortal combat with each other. Indeed, to defend the state was to defend the only existing moral constituency and there existed no superordinate standard in terms of which the means of waging war to defend this moral constituency might be pronounced unjust.

Machiavelli foreshadowed these major changes and it is within such a framework that his conception of war was rooted. Properly conceived, war has the purpose of creating, maintaining, and expanding the political order of the state. Adeptness in carrying out these political tasks is the measure of statesmanship and the supreme virtue in a world in which other virtues go unrecognized.

It does not follow from any of this that Machiavelli was an apostle of limitation in warfare. Indeed, such a supposition may appear counter-intuitive and has been hotly denied by some.[1] None the less, a plausible if not overwhelming case can be made for saying that Machiavelli tolerated certain principles of limitation in warfare in conformity with the political purpose of war: these principles were expressed in terms of a utilitarian means of reducing resistance (a promise of lenient post-war treatment could induce the enemy to desist), a doctrine of economy of force (military power is a scarce resource to be husbanded by the wise prince), and a principle of political arithmetic (in which it was necessary to avoid such overweening ambition as would fail to take account of the political realities and would lead to over-extension).

Whether or not Machiavelli advocated limitations in the waging of war as anything other than prudent and contingent, political manage-ment, what we can say with absolute assurance is that his is one of the finest examples of the political philosophy of war. War has its beginnings and its ends in the practice of the state and its relationship to other states. To this extent, it is autonomous and observes its own laws. But these laws are the same laws as those of politics more generally. Unlike Midlarsky, Machiavelli does not countenance a sharp break between politics and the use of force in war: both are manifestations of the same activity. It is not without passing interest that whereas later writers were to seek to constrain the practice of war by insisting that it was an extension of politics, Machiavelli's integration of war and politics is by far the more radical. It emerges from his imagery that politics is to be conceived of as no less than an extension of warfare. The language in which he describes the politics of statecraft is taken from the military language of war. It is therefore no coincidence that Machiavelli deems the realm of war to be governed by the demands of necessity, just as is the realm of politics, but it is the former which comes first in the logic of his system. We are therefore to admire, Machiavelli urges us, the heroic overcoming of *Fortuna* by the bold statesman, just as we would admire the subjugation of that lady by the inspired military commander. The rhetoric of politics and the rhetoric of war are one and the same but it is the former which is derived from the latter. Clausewitz was later to insist on the same intimate relationship but he was to seek to turn the Machiavellian universe right side up.

[1] For an elaboration of these points on Machiavelli, see I. Clark, *Limited Nuclear War* (Princeton University Press, 1982), 70–4.

Clausewitz

It is a judgement of a recent work that Clausewitz's lack of appeal can be accounted for by his being too practical for philosophers and too theoretical for military historians.[2] It is this achievement, of having produced a work of practical philosophy, that is to Clausewitz's greatest credit and that makes him such an admirable model for any study of the philosophical basis of the conduct of war.

If the argument thus far has been that it is possible to draw certain conclusions about the proper way of waging war from the concept of war itself, it would appear that the position adopted by Clausewitz in his celebrated work *On War* is an exception to this general rule. In the case of Clausewitz, if we are to understand what he is saying about the way in which war should be conducted, we must begin by ignoring what he says about the concept of war in its pure form. Rather than deduce conduct from the 'ideal', we can best understand war by looking as its 'real' embodiment. As Aron has remarked 'there is no absolute war in reality, it only exists in the world of concepts of ideals.'[3]

According to the ideal, the purpose of war is to impose your will upon your opponent and logically there would seem to be no reason to restrain the nature or degree of force applied in the pursuit of that end. On the contrary, in its ideal form, war will always, according to Clausewitz, tend towards its extreme and absolute form as a result of reciprocal actions between the two protagonists. In such a spiral of exertions, as each tries to overwhelm the other, there can be no place for restraint as this is contrary to the object for which the war has been undertaken. 'To introduce the principle of moderation into the theory of war itself', Clausewitz suggested, 'would always lead to logical absurdity.' This, however, is his first word on the subject and by no means his last.

War in reality is not like war in its idealized and absolute form. Hence, Clausewitz recommends instead that 'as soon as preparations for a war begin, the world of reality takes over from the world of abstract thought . . . and . . the interaction of the two sides tends to fall short of maximum effort'. It does so partly for practical reasons in that 'friction' intervenes and the sheer physical task of mobilizing and deploying forces in the field itself prevents the application of the

---

[2] M. Handel, 'Introduction', in Handel (ed.), *Clausewitz and Modern Strategy* (Frank Cass, 1986), 1.

[3] R. Aron, *Clausewitz: Philosopher of War* (Routledge & Kegan Paul, 1983), 69.

maximum degree of force of which the sides are theoretically capable. The mobilization and deployment of military power is a practical, not an ideal, activity and is governed by human qualities, physical obstacles, and elemental impediments.

The operation of these factors would, however, be purely random and the nature and extent of fighting, to a writer concerned to systematize the subject of warfare, could not be left in such a contingent state. Accordingly, Clausewitz went further to develop the theme that the application of force is moderated also by a more fundamental consideration. It is at this point that Clausewitz gives expression to his principal, if not consistent, conclusion that war is the continuation of politics by other means and is to be tempered by the demands of the very politics which have given rise to it in the first place. By this route we reach a second, and equally important, concept of war. 'Violence continues to be the essence, the regulatory idea, even of limited wars fought for limited ends', as Paret has argued, 'but in such cases the essence does not require its fullest expression. The concept of absolute war and the concept of limited war together form the dual nature of war.'[4]

This political philosophy of war is succinctly stated in his insistence that 'the political object will thus determine both the military objective to be reached and the amount of effort it requires'. A major practical consequence for the waging of war flows from this assertion. It means, in effect, that war will not develop to its natural extreme if your political objective is more limited because, in this event, you will not be prepared to expend such effort in its attainment. 'The more modest your own political aim', is his advice, 'the less reluctantly you will abandon it if you must.' Moreover, what might well lead to the abandonment of the goal is the realization that the costs entailed by the prosecution of the war exceed the benefit to be derived even from its successful termination. Accordingly, the analysis leads to its remorseless conclusion that '[o]nce the expenditure of effort exceeds the value of the political object, the object must be renounced'. In other words, the two protagonists are not involved in a mindless and mechanical process which propels them to extremes of violence but are instead engaged in a reasoned pursuit of their political objectives and determine how hard to try, and when to desist, in the light of the goals that are being sought.

[4] P. Paret, 'Clausewitz', in Paret (ed.), *Makers of Modern Strategy* (Princeton University Press, 1986), 198.

The energy expended in war will thus be modulated by the political intelligence directing it. War is not, as Aron explains,[5] like an explosion where the energy released is uncontrolled beyond the physical constituents at its point of detonation. War is located in a real political universe and is shaped both by its original causes and also by its continuing goals: 'Clausewitz precisely wants to prove that one cannot and should not separate a real war from its origins and ends.'[6]

Thus far the argument is relatively unproblematic. What remains uncertain is the status of this injunction. Having distinguished between war as an ideal, and war as it is practised, Clausewitz does not make entirely clear whether his political analysis of war applies only as an ideal, or whether he believes that this is how wars would be actually conducted. If it is true that friction can intervene, can it not also have the effect of liberating war from the political laws which should govern it, but might not always succeed in doing so?

Other matters of theory intervene at this point. How is the necessary 'guiding intelligence' of war to be maintained? More basically, perhaps, how can we be confident that policy seeks limited goals? Clausewitz's rudimentary injunctions about political control tell us little about the myriad complexities of the management of contemporary war, nor of the subtleties of civil–military relations. Many of the pressing technological circumstances which reinforce the need for political control of war make that task more difficult.[7] Handel is surely correct in insisting that Clausewitz 'could not foresee the complications of civil–military relations and their impact on the political supremacy necessary to the conduct of war'.[8] Nor does his formula address the corruption of politics. It avails us little to insist on the paramountcy of politics as a guide to war if that politics is tainted at source: Hitler's delict is not to be found in his failure to provide political leadership. These are judgements, of course, which Clausewitz was determined to eschew in his acceptance of war as political fact. But it is surely inadequate to make war an epiphenomenon of politics while leaving the latter unexplored, let alone explained.

Whether Clausewitz is engaging in description, or prescription, is not fully clear but the conclusion remains that either from a practical concept of war, or from its idealized version, we are counselled to

[5] Aron, *Clausewitz*, p. 66.   [6] Ibid. 62.

[7] See P. Bracken, *The Command and Control of Nuclear Forces* (Yale University Press, 1983).

[8] Handel, *Clausewitz and Modern Strategy*, p. 76.

make our military efforts proportional to the goals that we seek. Thus expressed, the political philosophy of war lends itself automatically to adaptation as a principle of limitation in warfare. The means of waging war will be limited in proportion as the ends of the war are themselves limited.

There is the rub. What will happen to the conduct of war if the ends arc unlimited? To say that war is limited by the requirements of policy provides no assurance in practice that the ends will not be all-consuming nor that political proportionality by itself can guarantee a war waged with restraint. To enunciate that war is a function of political policy may simply be to invite an infinite regression. Since the history of warfare is replete with instances of the costly means of war leading to an upwards revision of the aims of the war (the First World War became a war 'to end all wars' but did not thus start out), it may well be that the objectives of war are themselves, in practice, dependent upon its course and the manner in which it is conducted. States, in other words, have displayed a tendency to rationalize their losses in war by expanding their war aims, thereby retrospectively justifying the losses that have already been sustained. Theoretically speaking, this presents the possibility of another form of reciprocal action beyond that of which Clausewitz wrote—namely, that between the ends and the means both, in vicious circularity, driving the other to the extreme.

The problem did not escape Clausewitz's attention. He addressed it in his remark that 'as policy becomes more ambitious and vigorous, so will war, and this may reach the point where war attains its absolute form'. If the ultimate survival of the state is at stake, presumably there can be no resistance from policy alone to stem the rising tide of battle.

Post-1945 Theories of Limited War

What has primarily contributed to the emergence of a self-conscious body of writing about the subject of limited war in the post-1945 period has been the realization that war has lost most of its pre-existing 'natural' and 'political' limitations. The natural restraints upon the degree of violence that can be inflicted in the course of war have been systematically eroded by developments in military technology to the point where entire populations can be destroyed without the constraints of 'friction'. Similarly where once war was regarded, and certainly so during the century prior to that in which Clausewitz wrote, as the tool of dynastic monarchs seeking to adjust their territorial possessions at the margin, and eager to husband their resources in doing so, the political revolutions of the past two hundred years have liberated untapped political resources and allowed them to be mobilized in the pursuit of new national and ideological crusades. Again it must be reiterated that the practice of war owes much to the political theories which have lent it sustenance.

Given absolute weapons, coupled with the pursuit of potentially absolute objectives, war in the twentieth century has moved closer to its idealized form. Moreover, since war would not be limited by natural friction nor by political context, it was perceived as all the more necessary, if it was to retain some semblance of political purpose, that political artifice take up where nature had left off: if once limited war had been a free gift of nature, it had now to be the fruit of conscious political choice. It is certainly not the case that the idea of limited war has emerged only in the post-1945 period but it is equally true that no previous age has devoted as much intellectual effort as the present to what has been described as the 'deliberate hobbling' of existing military power. If hitherto the natural limits of war were tolerated as being in broad conformity with the value of the political goals being pursued, it would now be more accurate to say that limits in war are designed by political artifice.

As we have already seen, just war doctrine has been a broad church that has embraced a diversity of philosophical positions and arguments. Likewise, limited war theory has taken a number of distinct forms and, before we proceed to trace the outline of its development in recent decades, it may be useful at the outset to establish a number of categories to delineate both the forms which limited wars might take and also the sundry motivations which may underlie the enterprise. Some assessment of the distinctive nature of the motivations for adopting a limited war posture will be essential before a comparison with the philosophical bases of just war doctrine can properly be undertaken.

According to generally accepted categorizations, wars can be limited in a number of distinct ways.[9] They can be limited in terms of the objectives that are sought, the geographic scope of the conflict, the

---

[9] See, for instance, J. Baylis et al., *Contemporary Strategy* (Croom Helm, 1975). 'Guidelines' for limited war are summarized in W. V. O'Brien, *The Conduct of Just and Limited War* (Praeger, 1981), 222–3.

weapons employed, and, finally, the targets against which hostilities are directed.

Arguably, the first of these limitations is the most important one because, in the absence of restraint on the objectives of the war, it is difficult to see how other forms of limitation can be made to hold. Accordingly, we might distinguish between the objective of unconditional surrender which was the Allied war aim in the Second World War and the aim of Britain in the Falklands War which was to restore sovereign possession over the islands. It was not to overthrow the Argentinian government nor to restructure Argentinian society. To that extent, the objective was a limited one.

As regards the geographical scope of warfare, this has been a historically common form of limitation. Again, the case of the Falklands War illustrates its operation. Hostilities were confined to a maritime zone surrounding the islands and no direct hostilities were launched against the homelands of the two protagonists. It therefore provides a clear instance of a war limited in territorial scope.

Wars may also be limited by the choice of the weapons employed in waging them. Where a weapon is available for use, but a conscious decision is taken that it will not be used, we have an example of this third form of limitation. Inasmuch as Britain possessed nuclear weapons in 1982, but refrained from using them, the Falklands War is an instance of limitation in the means of war. Likewise, although gas was available to the belligerents in the Second World War, and despite the fact that its use in the First World War had created a widespread expectation that it would be used again, gas was not employed during the war.

Finally, wars can be limited by the nature of the targets attacked. In some cases this comes close to a restatement of the geographical scope of conflict as when hostilities are confined to targets in the field of battle and not elsewhere. However, when efforts are made to distinguish, for instance, between military and civilian targets, on or close to the zone of battle, it may be said that a fourth limitation is in operation. In the Gulf War, Iran and Iraq have periodically bombed each other's cities and periodically desisted from doing so. To the extent that the latter has been a matter of choice, and not strict military necessity, it has been presumably based on some calculation of the wisdom of refraining from attacking civilian centres.

Such an analysis of limited war provides us with some understanding of its nature. But this understanding is partial at best. All the above forms of limited war might be explained in strict conformity with principles deduced from doctrines of just war. If a tradition of limited war exists, and is thought to have some separate existence from the just war tradition, then the basis of its individuality cannot be found in the forms that the limitation might take but must be sought elsewhere in the motivations which lead to the implementation of those limitations. Accordingly, as an introduction to the understanding of limited war theory, we must seek also to construct some categories in terms of which limits in war are desired.

It might thus be suggested that theories of limited war have been concerned with a variety of possible motivations for attempting to restrain hostilities within certain limits. There is, however, an additional complication in recent theories of limited war in so far as some of the doctrines are intended for waging war in a controlled manner whereas others are interested in exploiting limitations as a means of avoiding resort to war altogether, by heightening the credibility of the deterrent response. Limited war doctrine has therefore performed the dual functions of 'war-prevention' and of 'war-fighting'. Unfortunately the line between these two positions is frequently blurred and, more often than not, is intentionally so. This complicates our analysis of the motivations behind limited war ideas, as the person who is concerned to prevent the outbreak of war entirely is not thinking along the same lines as the person who is concerned with what happens should deterrence fail and war occur. With this in mind, we might consider the following four positions:

1. Limitation may be sought, for utilitarian reasons, in the expectation of inducing reciprocity on the part of the enemy. If both parties bomb each other's cities, both may be worse off than they might otherwise have been without any material advantage resulting from doing so. From this perspective, it simply makes sense to forswear practices that are not militarily necessary provided that the enemy will reciprocate.

2. Limitations may be projected into a future war with the intention of persuading the enemy that the deterrent threat will be executed. In terms of credibility, a threat of massive reprisal may not be convincing to the enemy and may therefore fail in deterring him from a course of action. A threat of limited retaliation, coupled with a military capability to execute such a limited threat, may seem more credible to the enemy

as well as overcoming the problem of self-deterrence which might itself lead to paralysis of will. Thus in the context of a theory of deterrence, limited war may have attractions as holding out to the enemy a penalty which is smaller but more certain to be executed.

3. The third possible motivation is that which derives from the classical statement found in Clausewitz, namely, that the scale and nature of the war are to be determined by the value of the objectives being sought. It is a straightforward example of political accounting in which costs and benefits have to be entered on the balance sheet. However, the relationship between ends and means operates in both directions. It is not sufficient that the means be scaled only in proportion to the nature of the ends: in an age when the means of war have a capacity to outrun any intelligible political objective, the relationship must be inverted and the ends of war must be scaled down or abandoned if there is a danger of disproportionate means being employed.

4. Finally, limitations in war may be viewed simply as chips in a psychological bargaining process between the belligerents. A limitation is held out as a test of the opponent's will and nerve to continue with the struggle in the certain knowledge that to stay in the game holds out the prospect of this particular limitation being discarded and the struggle settling in at a new, and higher, level of violence. This entails a change in our perspective. We approach limitations in such a bargaining game, not by looking down at the degree of violence which has been held in check, but by looking upwards at the escalatory prospects and at the degree of violence still held in reserve.

The foregoing describes some of the intellectual apparatus which has been employed in the development of recent theories of limited war. They have ranged across the various forms and motivations for limitation already presented. It may be fair to suggest that what has given post-1945 theories of limited war their distinctive quality has been the existence of nuclear weapons in the strategic background. As a result of this, some would have it that recent theories of limited war are *sui generis*. For instance, one prominent theorist of limited war has contended that:

The detailed elaboration of a strategic doctrine of limited war, the formulation of specific plans for carrying out this doctrine, and the combined efforts of government, the military establishment, and private analysts and publicists to translate the doctrine into particular weapons and forces are developments peculiar to the nuclear age.[10]

The experience of the Korean War was to be critical in promoting the development of limited war theories. Its effects, however, were to come in two waves. In the first, the experience confirmed that, even in the nuclear age, limited wars were indeed possible. The Korean War seemed to be testimony that wars could be limited to specific geographical theatres, that they could be fought without recourse to nuclear weapons, and that, even when one of the belligerents was a client state of a superpower, such a war need not result in global war between the superpowers themselves. For these various reasons, the war seemed to validate the emerging doctrine of limited war.

None the less, Korea was an ambivalent experience and many emphasized its negative aspect. The fact that it had broken out in the first place suggested a failure of the West's posture of deterrence and the war, though limited, was costly and unpopular. In the short term, therefore, Korea pressed strategic theory in the opposite direction of total deterrence, encapsulated in the posture of massive retaliation. It was only in the longer term, as analysts came to question the credibility of this posture, that the second wave of the Korean War's influence broke upon doctrines of limited war. What was needed was a deterrent posture, and a military capability, both to avoid small-scale local wars and to prosecute them successfully should they break out.

In the early versions of this doctrine, the critical limitation was not deemed to be the nuclear threshold. On the contrary, Western analysts believed that to adopt a purely conventional strategy was to allow the Communist powers to exploit the advantage which they enjoyed in manpower and conventional forces. Militarily, therefore, it made sense to exploit the areas of Western advantage and they were deemed to lie in the use of tactical nuclear weapons. Accordingly, early scenarios of limited war were predicated on the use of tactical nuclear weapons, with full use of strategic hydrogen weapons threatened as a last resort. It was a major part of the theory, as articulated by the proponents of graduated deterrence, that such limited use of nuclear weapons could take place without necessarily triggering full strategic nuclear war between the contestants.

Such a distinction was to be proposed in an unlikely quarter in 1956.

---

[10] R. Osgood, 'The Reappraisal of Limited War', in Institute for Strategic Studies, *Problems of Modern Strategy* (Chatto & Windus, 1970), 94.

In the recently declassified British Cabinet papers of that year is to be discovered correspondence between the Archbishop of Canterbury and Prime Minister Anthony Eden.[11] The Archbishop lent his support to some of the contentions of the theorists of 'graduated deterrence':

You are familiar with proposals for distinguishing between the strategic use and the tactical use of nuclear weapons. . . . If we said that in any local war, we would use limited atom weapons up to an x degree of strength and no more unless and until our opponents went above that degree, a secure position is gained.

The virtue of such an approach, according to the Archbishop, was that the 'situation would thus come under moral control once again'. The argument was to be rejected by Eden, on both political and technical grounds:

No government could in my view define precisely in advance the circumstances in which it would use some weapons but not others, and any attempt at delimitation in advance would, I am sure, not be to the advantage of the West. Nor is it scientifically possible any longer to draw any clear line of distinction between the various types of nuclear weapons; the spectrum is continuous between the smallest weapon of kiloton yield and the largest with a yield of tens of megatons.

We have here a clear dialogue between two languages of limited war. The one asserts the need for limitation in moral terms, and in accord with moral purposes; the other retorts in terms of the requirements of effective deterrence and the technical state of the military art. Generally, however, and especially in the United States, the belief that gradations of nuclear war were possible was to be widely supported around the middle of the decade.

By the late 1950s confidence that this might be so had diminished. For a number of pragmatic reasons there was a tendency to doubt the efficacy of a strategy based on a threshold between tactical and other nuclear weapons. In the first place, exercises had produced the alarming information that a defence of Western Europe, in which tactical nuclear weapons were employed, would lead to such enormous casualties as to scarcely qualify for the appellation of limited war. Secondly, as the West began to lose such technological advantage as it had enjoyed in the production of these weapons, a strategy based on exploiting this advantage came to be increasingly suspect. Thirdly, the

---

[11] Public Record Office, CAB 129/79, correspondence dated 23 Jan., 22 Feb. 1956.

credibility of maintaining the artificial distinction between tactical and strategic use of nuclear weapons persuaded some theorists, including Henry Kissinger who had initially promulgated the virtues of tactical nuclear defence, that the only tenable and salient threshold in contemporary conditions of warfare was that between nuclear and conventional weapons. Hence, by the early 1960s, some augmentation of Western conventional forces was the order of the day, though it was still coupled to tactical and strategic deterrence via the menu of flexible response.

There also emerged a tendency to discuss limitation within the actual context of an ongoing strategic exchange and this perspective was to become of increasing importance. Whereas there had been an early reluctance to 'think about the unthinkable' and to contemplate the nature and possible course of a nuclear war, some bold spirits began to suggest that even a strategic nuclear war might observe its own pattern of limitations. These might, most obviously, take the form of restrictions on targets with some attempt to discriminate between countervalue and counterforce objects of attack. Additionally, there might be restrictions on the overall scale of attack (at least initially) as well as upon the rapidity of the sequence of hostilities. Even with a nuclear war there might be time for the bargaining process of mutual intimidation to work itself out and for the task of politics to be performed: the function of limited war theory at this point becomes, not the avoidance of war, but the deterrence of a larger-scale war than the one in which the contestants are already engaged.

The general course of limited wars of the post-1945 period has not been such as to inspire much affection for this particular genre of warfare. The Vietnam experience served only to heighten doubts about the validity of this kind of military enterprise. But the reason why limited war theory has persisted is presumably the same reason as inspired it in the first place, and that is that whatever the deficiencies of limited wars, they none the less remain more attractive than their unlimited alternatives.

Just and Limited War Compared

How do the just war and the limited war traditions relate to each other? Are they opposed or mutually compatible traditions of warfare? As James Johnson has asked:

What is the relation of the ideas making up the concept of limited war to those defining just war as we know it? Are these better understood as two separate traditions on restraining war within Western culture, or rather as two types of emphasis within one broad tradition?[12]

In the following section, an attempt will be made to demonstrate the points of similarity between just and limited war, and this will be followed by a review of the points of contrast. This should facilitate the endeavour to answer the above question and to see the full relationship between the two doctrines. Indeed, it may be possible to shed some light on the interpretation that limited war is an expression of the perennial search for a more civilized form of warfare but is the farthest that twentieth-century conditions permit us to go in realizing that quest. This in itself would be an unconventional wisdom. If anything, the more orthodox view would be that restraints in the conduct of war deriving from political considerations are fundamentally different from those which have their basis in just war principles. As Peter Paret observes in a recent essay on Clausewitz, he 'is far closer to Machiavelli's position than to that of the Church fathers and of moral philosophers who want to define the just war and just behavior in war'.[13] How wide, in fact, is the gulf between the two traditions?

Similarities between Just and Limited War

1. As has already been noted, there is an identity of form as between just and limited war traditions. The just war notion of the need for a just cause intended to restore a just peace, and of the gross limits which the *ius in bello* imposes upon the conduct of war, finds its counterpart in the notion of limited wars which can be limited, most fundamentally of all, in terms of the objectives sought, but also by way of restrictions on the scope of hostilities, the targets to be attacked, and the instruments of war deemed appropriate to the purpose.

2. The notion of proper authority appears in both traditions. In accordance with just war precepts, a necessary condition of a just war is that it is undertaken by a duly constituted and legitimate authority, thereby excluding from its provenance those other forms of violence which do not conform to this political ideal. In essence, this

[12] J. T. Johnson, *Just War Tradition and the Restraint of War* (Princeton University Press, 1981), 190.
[13] Paret, *Makers of Modern Strategy*, p. 210.

requirement has served to forge a bond between our conception of war and our conception of the political purpose of the state since no other form of proper authority has been duly recognized.

In similar fashion, albeit with differing emphases, the limited war tradition, deriving from the assumptions of a political philosophy of war, has associated the intrinsic nature of war with the purposeful activity of the national state. In order that war might follow its essential nature, it follows also that the direction of the war be in the hands of the duly constituted political authorities. It is not a matter to be turned over to military professionals, however necessary their technical advice might be. Accordingly, it has been a perennial theme of the limited war tradition, at least since Clausewitz, but equally prominent in all post-1945 versions of limited war, that any war can remain true to its political purpose only if it remain subject to political control and direction by the proper authorities.

3. Both traditions, albeit with some necessary qualifications, have consciously related their concept of war to a code for its prosecution. In neither case has this code been presented as a set of absolute and unchanging prohibitions but in both cases there has been some philosophical attempt to derive principles for the waging of war from a generalized concept or statement about the nature of war itself. The point can be made by claiming that *ius in bello* restrictions do not stand on their own as a set of general moral precepts because, without some frame of reference, they are superfluous. Their significance is derived from their nature as a statement about what may permissibly be done even in a state of war and it is consequently imperative that the frame of reference be clearly delineated.

In the same way, the political precepts of the limited war tradition have no meaning outside the political concept of war which is their point of origin. Without some purpose to military activities, as Clausewitz suggested, war would tend towards its theoretical extreme. There can therefore be no point to limiting the means of war less it be in association with some concept of what the point of the war itself is understood to be.

4. Paradoxical as it might at first sight appear, both traditions make direct appeal to universal standards in establishing their respective principles of conduct. Just war tradition has been grounded on a variety of divine, natural law, and universal human right assumptions. Inasmuch as it has aspired to establish a code of conduct of universal validity, surpassing individual cultural and historical circumstance

(without, of course, necessarily succeeding in doing so), it has been necessary to relate those precepts to some universalist philosophy in terms of which the precepts might have a general validity. Indeed, those periods during which just war ideas have most noticeably faltered have been precisely those historical periods when universalist philosophies themselves were under most direct challenge.

The same comment is less apparent in connection with the limited war tradition. Can we speak of universal norms as the underpinning of such ideas? It is perhaps more difficult to give specific content to the philosophical assumptions associated with limited war theories, but in so far as they seem to have an expectation that states of widely differing cultural persuasions and political forms might none the less adapt themselves to limited war practices, there can be little doubt that universalism, even if covert, is at work here. What the tradition seems to appeal to is some cosmopolitan political or prudential standards that all states will recognize and allow them to conform to specific limitations in the conduct of war. For instance, despite some writings which have argued that Soviet perceptions of nuclear war are culturally distinct, there has been a much more widespread tendency to assume that all states have broadly similar attitudes to the uses, and limitations upon use, of nuclear weapons. The very abstract theoretical discussions of nuclear deterrence seem certainly to have proceeded on some such assumption. If there is indeed a contemporary right-reasoning *homo strategicus*, might it not be said that we have reached a new form of cosmopolitan and secular naturalism?

5. It needs to be emphasized that, by definition, the stress of both traditions is upon wars as being fightable. It is perfectly possible to depict both just and limited war theories as a means of deterring wars but it would not be possible to present them as being philosophically opposed to the actual fighting of war. The theoretical point of departure for just war doctrine is that it is a right, and in some circumstances a duty, to take up arms in anger. This was demonstrated from the beginning in the rejection by just war theorists of the pacifism of the early Christian Church.

In the same way, even though much of the theoretical intention associated with the development of limited war ideas has been the better to prevent war's outbreak, limited war doctrine is also philosophically divorced from pacifism and from other creeds of abolition. Limited war doctrines have been promoted precisely because they make wars more fightable (either in practice or as a means of more credible deterrence) or because, if wars should none the less occur despite our best endeavours, it makes sense to fight them in such a way that the resultant damage is minimized. Such a perspective is significantly different from that which would renounce war in all circumstances. In other words, both just war and limited war traditions are, quite literally, philosophies for the waging of war.

Contrasts between Just and Limited War

For these various reasons there are points of identity or similarity between the two traditions. None the less, there remains a feeling that the two should not be equated and that there should be some separation between them. For instance, while it may be appropriate to suggest that Clausewitz is an exponent of politically limited warfare, there are few who would feel comfortable to have him presented as a proponent of just war. What then might be the points of separation between the two traditions?

1. While it has been suggested above that both the just and limited war traditions can be adapted to purposes of deterrence, it has to be said that limited war lends itself more readily, and to a greater extent, in this direction. It may well be that the knowledge that just wars will be fought may act to prevent unjust aggression, but just war can never be a doctrine of absolute deterrence. By way of contrast, some of the theories of limited nuclear war tend towards the pole of deterrence and at that point part company with the just war tradition. The reason that some strategies of limited nuclear war are recommended is that, given their greater credibility, there is less likelihood of a need to resort to their execution.

However, a situation of absolute deterrence which prevents all wars would also prevent those very just wars that are needed to rectify an unjust situation. Accordingly, inasmuch as there is a right, and in some circumstances a duty, to just war, it cannot be sacrificed on the altar of war prevention. Thus just war distances itself from absolute deterrence as much as it does from other creeds for the renunciation of war.

2. There is also a sense in which the limits of limited war are to be regarded as inherently more flexible than the limits of just war. This is a statement of degree and not of kind. It has already been argued that the precepts for the conduct of war established by just war argument are seldom, if ever, absolute ones. The tendency of the tradition has

been to create a relationship between moral principle and the requirements of military necessity rather than to suggest that the latter be completely overridden by the former. The principle of double effect is but one example of the attempt to create such a balanced relationship.

None the less, it is difficult to discern the philosophical content of any intrinsic limits in the limited war tradition. Since the motivations underlying the adoption of limits may themselves vary, so will the nature of the limits themselves. If the intent of the limits be but to induce the enemy to reciprocal respect for prohibition of certain methods, the use of which is to neither side's advantage and the prevention of which is in mutual self-interest, then it may still be the case that, if certain limits are infringed, there persists an interest in restricting hostilities at the new, albeit higher, level. There may in that circumstance be more inconvenience but it is not at all clear that some vital principle has been sacrificed.

This is equally apparent in the case of limitation undertaken in accordance with the general precept of determining the means of war in correspondence with the political objectives for which the war is being fought. To constrain thus the means of waging war is to operate on the basis of a sliding scale which permits the intensification of military conduct in proportion as the ends are valued. Again there is little philosophical content in the individual limitations themselves beyond the gross limitation upon undertaking measures which violate the proportions of war. This may possibly restrain such measures beyond which no intelligible objectives can be achieved, but it has to be said that the code is utterly permissive of the nature of all violence up to such a point, if it can ever be recognized in practice.

The same is even more conspicuous in the case of limitations viewed as part of the psychological bargaining process with the enemy. If the function of limitations is to hold out the threat of increased escalation, and if it matters little whether the opponent yields now or later, as long as yield he does, the individual limits of war are devoid of intrinsic philosophic content. They serve merely as bargaining chips and not as statements of principle.

3. Perhaps the simplest way to distinguish between the two traditions is to suggest that limited war belongs to the realm of politics whereas just war belongs to the realm of morality. This position is beguiling in its sweet simplicity. But its compelling straightforwardness is also the source of its baleful inadequacy. None the less, the idea is of

sufficient merit to be worth using as a point of philosophical departure. It is certainly true that most of the argument about just war is couched in the language of moral absolutism whilst the limited war debate is conducted in the more tempered dialect of political instrumentality and prudence.

Beyond this point, however, such a sharp distinction is of little avail. The issue has been well highlighted by Stanley Hoffmann:

> The attempt to moralize interstate conflict cannot consist only of legal and ethical rules for, or in, war; it is inseparable from the political game. . . . Once again we find that we should not pose the problem as a problem . . . of morality vs politics. It is through the right kind of politics that some moral restraints can become observed and practicable.[14]

Hoffmann's argument serves well to soften the edges of morality and politics and to show that the two can indeed be integrated. Unhappily, as a means of resolving the present issue, far from simplifying the matter, it complicates it beyond measure by greatly extending the terms of reference. What the argument does is to compel us to debate the same issue in a yet wider context by investigating not only the nature of just war, and the means proper to it, but also the very nature of just politics.

4. A fourth possible mode of distinguishing between the two doctrines is that suggested by James Johnson in response to his own question. Be it noted, however, that his final position is that, while there are ways of distinguishing between just and limited war, the two coexist *within* a common tradition.

The difference between them, then, in Johnson's analysis, is that limited war gives greater emphasis to proportionality over discrimination while the reverse is true of just war. The stress in limited war is upon reduction of the total amount of destruction whereas the stress in just war is upon the protection of non-combatants. He provides his own illustrative example to demonstrate the difference between the two:

> Their fundamental distinctiveness can be grasped readily by conceiving of a hypothetical war in which a sudden and terrible slaughter of noncombatants brought an abrupt end to the fighting, as opposed to a war in which successful efforts to protect noncombatants prolonged the destruction of a country's economic and social base.[15]

The point of this argument is lost because of the extent to which it is

---

[14] S. Hoffman, *Duties Beyond Borders* (Syracuse University Press, 1981).
[15] Johnson, *Just War Tradition*, p. 208.

overstated. If the just war tradition means anything, it surely is a philosophical combination of both proportionality and discrimination, and the same might be said of limited war. By presenting it thus, Johnson distorts both traditions by turning them into caricatures, each with a unidimensional obsession. It is not at all clear that the former alternative would be in conformity with limited war, nor indeed that the latter alternative would conform to the requirements of just war.

5. A final, and intriguing, mode of drawing a distinction between the two traditions has been suggested by Paskins, even if for another purpose:

Many see the just war ideas as a worthy effort at limiting war: war is something given and the task is to do what one can to mitigate its disasters. But this is to begin at the wrong end. The problem is not how far war can be limited, but how far it can in conscience be extended.[16]

The idea is suggestive. What it holds out is the possibility that, even should the two traditions converge in the centre, they continue to differ on the basis of their philosophical points of departure. Limited war, as it were, starts from the perimeter and seeks to push the limits of war inwards, thus delineating its area of proper conduct. Just war begins at the core and seeks to push the limits of war outwards, defining the legitimate reach of battle. The former discusses the area of required restriction: the latter discusses the area of necessary permission. The implications of this distinction will be further considered in a later section of this book.

The preceding chapters have outlined two broad traditions in terms of which the concept of war might be understood. There are extensive areas of overlap between the two—at least as regards substantive results, if not by way of the reasons underlying them. In O'Brien's terms, limited war practices are a necessary, although not sufficient, precondition of just war.[17] There remains, however, adequate intellectual distance between the two for them to be distinguishable at the edges.

It is now time to move away from a consideration of general concepts and broad traditions to a more detailed treatment of specific codes of war and problems encountered in their implementation. This is where the philosophy of war confronts its second great test. It is one achievement to articulate comprehensive, compelling, and consistent

---

[16] B. Paskins and M. Dockrill, *The Ethics of War* (Duckworth, 1979), 229.
[17] O'Brien, *The Conduct of Just and Limited War*, p. 348.

principles of war: it is another to translate these principles meaningfully into the face of battle. Thus far this book has maintained that the idea of war is a powerful determinant of war's practice and that we make war in accordance with our image of it. It is now necessary to relax this contention and to proceed in a more qualified manner.

Our idea of war shapes the reality but imperfectly. As in the contention of Clausewitz, what separates absolute war from war in reality is the friction which intervenes. We shall not concern ourselves here with the same military forms of friction as preoccupied Clausewitz but the principle at work is similar. Whereas it is possible to elaborate general codes for the conduct of war from the broad concepts of war available to us, it is in the process of implementation that philosophic friction makes its presence felt. It is to this grey area that the analysis must now turn.

4

CONDUCT BECOMING AND UNBECOMING

THE regulation of warfare can occur in a number of differing ways and it will be the task of this chapter to analyse some of the most important of these. It will begin by discussing the nature of war crimes and the associated problem of responsibility for their perpetration. This point of departure is relatively straightforward, being concerned with the subject of inter-state war and the laws of war which govern it. From there the less than paradigmatic case of guerrilla or insurgency warfare will be examined with a view to exploring the difficulties entailed in applying the rules of war to a non-conventional military environment. Finally, the chapter will review the two principal modes for the restraint of war—restrictions on weapons and targets—and point to the sundry difficulties created by their implementation.

Crimes of War

War and the Idea of Crime

The course of twentieth-century warfare has applied intense pressure upon the laws of war. Various technological developments, principally in the realm of air power, as well as the scale and ferocity of military campaigning, have served to erode traditional distinctions thought to be crucial to the maintenance of conventions limiting the conduct of war. The enactment of crimes against humanity during the Nazi terror added yet another dimension to the horrors of war. Seen in this context Glover's comment gains some sympathy, however overstated it might be: 'It is ironic that the first international trials of war criminals should have started at the period when the rule of law in war had collapsed, leaving only vestigial remains.'[1]

The argument so far has been intended to demonstrate the

---

[1] M. Glover, *The Velvet Glove* (Hodder & Stoughton, 1982), 247.

relationship between concepts of war and the ensuing practices within them. Here it must be said that the idea of a war crime encapsulates an entire philosophy of war. To speak of a crime of war is already to make certain fundamental assumptions about the nature of war and about the acts which are permissible in its pursuit.

The point can be made by locating the idea of a war crime on a spectrum between the following extremities. At the one pole it would be contended that, given the nature of war, *all* acts undertaken in its name are fair because it is the absence of overarching standards of agreed justice between states which leads them to war in the first place. At the other extreme we find the idea that war itself is a crime and so must be all of its acts. If the purpose of war is criminal, the means of waging it share in the general indictment.

The notion of a crime of war is to be discovered somewhere between these extreme positions. It occupies the middle ground. In the same way as just war theory is a delicate blend of permission and restriction, so the doctrine of war crimes holds that there are indeed acts of war that are permissible which would not otherwise be so outside the setting of war, but that by no means all such acts meet the test of acceptability. Acts of war are neither to be condoned nor prohibited in the round: rather we must seek to establish the permissible realm of war. Whereas up to this point we have sought to relate concepts of war to resultant practices, we shall now concentrate upon the implementation of the general principles already derived. Since the doctrine of war crimes is restrictive and tells us what may not be done in war, it is equally permissive, even if only implicitly so, about the acts that it does not prohibit. The practice of war crimes is therefore incompatible with either of the extreme formulations of war—the extremity of permission implied in the 'war is hell' doctrine and the extremity of restriction implied in the doctrine that war itself is a crime.

In certain respects, however, the term 'war crime' is itself a misleading one. The mental picture which this conjures up is that there are crimes which are distinctive to the state of war. But this is scarcely so and tends to present the situation the wrong way round. The crimes of war would be equally crimes in peace and remain as crimes because they are not held to be legitimate acts of war. In other words, the legal state of war provides certain dispensations for the perpetrators of military acts to undertake actions which would in peace be illegal, provided that those acts are undertaken with due authority and in conformity with the legal requirements of the code of war. It is not war, as such, which makes them crimes. Unjustified killing is illegal and remains so in war because, unless the act conforms with the practices of war, war itself can provide no immunity for its perpetrators.

There are at least two broad approaches available in establishing the limits beyond which acts of war become crimes. These approaches underlie the doctrines of just and limited war and also the practical endeavours to create a body of positive international law for the regulation of armed conflict. According to the first approach, it is preferable, out of pragmatic respect for the exigencies of war as a practical activity, to set the bounds of permissible conduct as generously as possible but then to insist that it is absolutely forbidden to go beyond these limits in any circumstance. Its philosophic approach is lenient in terms of the content of the rules themselves but strict in terms of its demand for their observance. The second approach goes about its business of devising rules for the conduct of war in a philosophically distinct way. It defines the limits of permissible conduct in a restrictive manner but then allows that individual cases for going beyond them can be made if the circumstances are special. While it is strict on the rules themselves, it is relatively more lenient in allowing exceptions to them.

The issue is of some considerable practical import and bears on the related matter, already raised, of whether the function of the rules of war is to restrict or to extend the reach of battle. In accordance with the former approach there is substantially less scope for philosophical debate once the rules are agreed because what it does is to mark off a narrow band of activities as illegal and to suggest that crossing this line is an unambiguous step into criminal territory. The latter allows a small area at either extremity, that which is absolutely permissible and that which is absolutely impermissible, but concentrates its energies upon the middle ground where the debate takes place. Transgressing the boundaries becomes something which requires justification in the special circumstances.

This expresses the issue rather abstractly. The consequence of these differing approaches is that, in accordance with the greater liberality of the former, that which is not absolutely prohibited is absolutely permitted: an act is either permissible or not. The latter allows for a more subtle differentiation. In addition to the realms of permission and prohibition is the area of debate where what is normally prohibited can be argued for if the circumstances are sufficiently compelling. This may have the disadvantage of being too flexible in the face of the requirements of military necessity since almost all national leaders and military commanders can present their own circumstances as being unique. But it also has the benefit of allowing for a category of actions which are illegal in normal circumstances but possibly justifiable in the exception. The former approach tends to give blanket endorsement to this grey area; the latter to make it an area of philosophic debate where the onus is upon the actor who would go beyond the pale to demonstrate what the justification is for doing so.

The idea of war crimes, finally, touches upon the deepest questions of political philosophy concerning the nature of the state. As will be discussed below, these questions are posed also in the context of defining the proper targets of warfare and in the elaboration of principles of discrimination. For the moment, these selfsame issues are raised in the context of distinguishing who bears responsibility for acts of warfare. It is essentially a problem of disaggregation, of trying to locate the proper bearers of responsibility within a large corporate structure such as the state. Who, then, acts in the name of the state in waging war? Is it the supreme political authorities alone? Is it the senior military commanders? Is it the soldier in the front line? Or is the responsibility shared amongst every individual member of the warring state? At one level these are questions of moral and political philosophy. But if the idea of war crimes is to be embedded in a code of legal practice, and if the military codes of discipline of national armed forces are to take practical cognizance of them, there must be some tangible way of assigning and apportioning the responsibility for illegal acts of war among the persons who have acted in the state's name.

Crime and Rules of War

There can be no crime of war unless war is governed by legal rules. These rules stipulate how far the dispensation of war is to carry and what remains a crime because it is not covered by the category of legtimate acts of war. Part of the reason for the elaborate procedure which has normally been associated with the business of initiating war, even though the practice of formal declaration of war has fallen into relative disuse, is that it was necessary for the legal state of war to be affirmed and recognized in order that the rules appropriate to this condition might be implemented.

In practice, rules of war have been enunciated, and to a degree respected, for a large number of disparate reasons. It is possible to enumerate the following instances:

1. Mutual self-interest in avoiding unnecessary suffering or in outlawing unpredictable weapons of dubious military advantage.
2. Economic motives, as in the case of ransom, where lives are spared, but primarily for financial gain.
3. As an adjunct of maintaining discipline which is necessary for the efficient execution of military tasks.
4. By custom, since it is a commonly observed practice in any case.
5. In legal positivist terms, because it is a law.
6. Because they have moral appeal and strike a responsive chord within us.

The last argument contains within it, however, a crucial ambiguity which must be analysed for the sake of clarity. A critical distinction exists between a conception that there is moral value in having some kind of rules of war and the conception that the only moral value resides in the content of the rules themselves. For instance, if we believe in *any* rule which reduces the number of those killed in war, or which controls the nature of the suffering to which people are subjected, we might argue that any rule of warfare which has such an effect is morally valuable and a war fought in accordance with such rules is preferable to one which is not. Accordingly, in the light of this notion, it might be appropriate to introduce a rule of warfare, indeed a principle of discrimination, whereby only children were deemed to be proper targets of warfare. The conduct of war would therefore be restricted to activities directly against children. The great merit of this principle is that, in sum, a war fought by this rule would, of necessity, produce fewer casualties than a war fought without such restriction. The rule must therefore be morally desirable.

No sooner is this argument presented than we are led to deny its validity. There are a number of things which seem to be wrong about the adoption of this principle but, above all, the very content of the rule seems lacking in moral coherence. Not only is the rule morally arbitrary in selecting children as the targets of warfare but it is counter-intuitive in so far as the targets of warfare, on this reckoning, bear no logical relation to the purposes of war itself: if it is hard to see why the reach of war should be extended into this area, it is doubly difficult to see why it should be extended into this area alone.

If we are therefore talking about a moral basis for the rules of war, we are pressed to insist that, while the existence of some rules of warfare is necessary for the view of war as a moral activity, it is not a sufficient condition. What is also necessary is that the rules themselves have moral content and strike us as morally coherent, given the nature of war itself.

The legacy of the tribunals at the end of the Second World War has been such as to leave three categories of war crimes: crimes against peace, war crimes, and crimes against humanity. This last category goes beyond the confines of the discussion about conduct in war (hence the invention of this separate category) and will not be further discussed.

As to the crimes against peace, they may be seen as legal translations of violations of the *ad bellum* component of just war theory. What is prohibited by this is the 'planning, preparation, initiation, or waging of a war of aggression'. In short, the law of war is deemed to proscribe actions leading to the inception of that particular form of unjust war which is labelled as a war of aggression. To that extent, the Nuremberg and other tribunals were the culmination of that tendency towards the revival of just war doctrine in the twentieth century which had been anticipated in the Hague conferences, in the League Covenant, and in the Kellogg–Briand Pact.

War crimes proper may likewise be seen to be associated with the *in bello* component of just war and to focus upon the actual conduct of war once entered into. War crimes, positively defined, are then 'violations of the laws or customs of war'. Their provenance is the treatment of the civilian population, of prisoners of war, of hostages, etc.

Just as theorists of the just war have experienced difficulty in finding a proper accommodation between the twin components of their doctrine, so have legal theorists argued variously about the inter-relationship between the legality of the war itself and the legality of acts committed in its name. Do acts of war become criminal only to the extent of their own intrinsic violation of the laws of war? Or is their criminality influenced also by the illegality of the war in which they are undertaken. Some philosophers, such as Nagel, have expressed the view that 'if the participation of the United States in the Indo-Chinese war is entirely wrong to begin with, then that engagement is incapable of providing a justification for any measures taken in its pursuit'.[2] In essence, the structure of this argument is that since a crime remains a crime, unless or until its status is changed by the legal condition of war, all acts undertaken in pursuit of an illegal war remain criminal and cannot find exoneration. In this argument we have a striking example of the criminality of the conduct of war being derived from the criminal character of the war itself, regardless of whether the acts performed are in accordance with the substantive laws of war or not.

What is to be made of this line of reasoning? Might it not be objected that if the status of the war in *ad bellum* terms can have this capacity to render illegal acts of war that would otherwise be considered legal, this argument has the unhappy corollary of proposing also that the state which enjoys the greater justice, also enjoys the greater licence in its conduct of war as part of the fruits thereof. In other words, the justice of the cause must also logically have the power to convert illegal acts of war into legitimate ones. Does it not follow that if all acts are illegal in pursuit of an unjust war, then also all acts are legal in pursuit of a just one—a position sensibly rejected by international lawyers?[3]

It is not at all clear that this is a necessary conclusion to draw. Indeed, on the contrary, it can be reasonably objected that the justice of a war is contingent upon the manner of its conduct and is not a final condition. While there are good reasons for dissenting from Nagel's view by objecting that the participants in an unjust war do not lose their legal rights, there are even stronger reasons for distancing ourselves from the view that the participants in a just war acquire additional rights over and above those which they would normally enjoy as belligerents.

Responsibility

Political scientists have invested much time and effort in recent years in trying to discover the locus of decision-making in complex organizations. A similar problem confronts the legal application of the laws of warfare as there is a need, at some point, to settle the question of responsibility if the laws are to be enforced. Accordingly, at the end

---

[2] T. Nagel, 'War and Massacre', in M. Cohen *et al.* (eds.), *War and Moral Responsibility* (Princeton University Press, 1974), 3.

[3] See C. Greenwood, 'The Relationship between *ius ad bellum* and *ius in bello*', *Review of International Studies*, Oct. 1983.

of the Second World War, the International Military Tribunal pronounced its view that 'crimes against international law are committed by men, not by abstract entities, and only by punishing individuals who commit such crimes can the provisions of international law be enforced'. No longer would the state alone be held accountable before international law but the individuals who act in its name. This resolved, there remains the matter of distinguishing which individuals are responsible for crimes of war.

Assigning such responsibility is no easy matter. Judgements may vary depending upon whether it is crimes against peace (the policy issuing in war) or war crimes (conduct in the war) which are at issue. There is a general problem, depending upon the political system of the individual state, in deciding whether the political chain of command, or the military chain of command, is the appropriate body within which to begin the search, or, if a mixture of both, in which manner the two are to be combined. It is possible to address the question of responsibility in purely formal terms (equivalent to constitutional doctrines of ministerial responsibility) where the holder of an office is deemed to be at fault, regardless of the specific facts of the case, because part of the office is the acceptance of such responsibility. Walzer finds that in the case of the Japanese commander Yamashita, he was found guilty of war crimes on some such formal doctrine and, as Walzer comments critically, was 'convicted of having held an office'.[4] Whether formal position gives the incumbent the actual power to influence policy or conduct is a moot point and, accordingly, in discussing crimes against peace, the International Military Tribunal claimed its own position was one of behavioural responsibility rather than formal responsibility. In its own words, 'it is not a person's rank or status, but his power to shape or influence the policy of his state which is the relevant issue for determining his criminality under the charge of crimes against peace.'

Over and above the complexities of distinguishing the individuals who are to bear responsibility, is the closely associated problem of deciding what weight is to be attached to various pleas in mitigation. Of the many items which might be considered under this heading, three alone will be mentioned.

1. First, there is the vexed subject of superior orders. This is a plea that responsibility must be redistributed upwards within the chain of

<hr />

4 M. Walzer, *Just and Unjust Wars* (Basic Books, 1977), 320.

command. If the soldier is ordered to commit an act which is deemed to be in violation of the laws of war, is he responsible for his act if he was ordered to do it? How, indeed, is the soldier to make a determination on the legality of the order which he has been given having but a partial view of the field of battle and being incapable of judging whether a particular operation is strictly militarily necessary or not? Given the vagueness of the laws of war themselves, and the extent to which they are qualified to take account of military necessity, how can the individual soldier reach a reasoned judgement? Even if he could, is he a free agent? Presumably any failure to carry out an order is a breach of military discipline, carrying a possibly severe penalty. How far is the soldier required to go in incurring personal penalties (reprimand? execution?) in resistance to orders known to be in contravention of the laws of war? The Tribunal recognized this latter issue in proclaiming that 'the true test . . . is not the existence of the order, but whether moral choice was in fact possible'.

2. Additionally, it may be that, in determining criminality of individual acts of war, account must be taken of the general ethos in terms of which the war is directed. High-level political decisions may compel the armed forces to adopt certain strategic and tactical practices known to infringe the laws of war. It is to this notion that Roling makes appeal in elaborating his concept of 'system criminality'. 'The criminality depends on societal forces', he suggests, 'rather than on personal inclinations, the effect of these forces ranging from direct orders, through official favour, to conspicuous indifference. The crime is caused by the structure of the situation and the system, and might therefore be called system criminality.'[5] Might not the front-line soldier then make the same plea as the hapless messenger of old, that he should be spared execution as he is simply the bearer of bad news and not the creator of it?

The attitude of the law is not, however, so lenient. Being instructed in the laws of war, the soldier has an individual responsibility to question orders deriving from a general style of military operations known to be unlawful.

3. Finally, there remains the defence of reprisal. The fact that an enemy has also committed an offence against the laws of war does not make your own violation of the same laws any less illegal. But in so far

[5] B. V. A. Roling, 'Aspects of the Criminal Responsibility for Violations of the Laws of War', in A. Cassese (ed.), *The New Humanitarian Law of Armed Conflict* (Editoriale Scientifica, 1979), 203.

as the laws of war have customarily allowed a limited scope for reprisal, as a means of self-regulation in support of the laws themselves, the plea of *tu quoque* may be presented as a mitigating circumstance. Some retaliatory acts can fall within international legal requirements but there is certainly no unlimited licence to undertake such acts. Just as resort to war, in just war terms, is permissible in extremity when other recourses have been exhausted, so reprisal should be a final resort—'ultimate measures to be resorted to with a sense of moral responsibility when no other means to recall the guilty State to respect of law were available'.[6]

The Regulation of Irregular Warfare

The stipulation of what constitutes a crime within a clearly defined category of war is one problem in the implementation of codes of warfare. A second general issue is that of establishing the applicability of the laws of war to those armed conflicts that are deemed to be irregular in some important sense—guerrilla, revolutionary, insurgent, and terroristic warfare being the most commonly cited examples.

The regulation of warfare, and its conceptual 'regularization', go hand in hand. The attempt to constrain the practices of war is partly a conceptual task of clarifying what war is and how far its writ may legitimately run. In this sense a principal tendency has been for the laws of war to colonize—and properly so—those other armed conflicts which had not traditionally come under its purview. Had it not done so, and given the expansion of irregular warfare at the expense of conventional war, the legal regulation of armed conflict would have applied to a decreasing proportion of the violent political struggle taking place around the globe.

Two approaches are discernible in the attempt to regulate irregular warfare. The first is the tendency to make such struggle conform to the conceptual norm in order that the writ of existing law of war might be extended to it. This is what Johnson refers to when he describes, in relation to insurgent warfare, the emergence of a 'pressure to transform such conflict into the "normative" kind of warfare represented by conventional military clashes between established powers'.[7] The second is the assertion of the actuality, or the equal legitimacy, of other

[6] R. Bierzanek, 'Reprisals as a Means of Enforcing the Laws of Warfare: The Old and the New Law', ibid. 235.
[7] J. T. Johnson, *Can Modern War Be Just?* (Yale University Press, 1984), 54.

forms of armed struggle, and the search to develop meaningful restraint of violence in those other, and conceptually valid, forms.

A useful example of the latter can be found in one discussion of the necessity for belligerents to wear uniform and of the infringement of military conventions by those who do not. As we shall note below, it is a consistent objection to the forms of warfare currently under review that they fail to guarantee meaningful separation between combatants and non-combatants and the failure to wear a distinctive military tunic is but a symbolic representation of all that is 'evil'[8] about those modes of waging war. As against this, Fotion and Elfstrom castigate the conceptual limitations of this approach:

There is nothing wrong with first looking at war . . . by focusing on establishment inter-nation wars. That is, there is nothing wrong so long as one does not start thinking of these kinds of wars as paradigmatic and then supposing that other kinds such as civil and guerrilla wars are deviant. . . . [T]he immediate purposes of nations who may have legitimate reasons for engaging in war may not be the same as the immediate purposes of non-establishment groups who may have equally legitimate reasons for engaging in war.[9]

Historically, irregular war, and especially civil war, found itself largely outside the scope of the positive laws of warfare, so much so that O'Brien complains of this as a failure which has 'left a normative gap in a most critical area of modern conflict'.[10] There were many reasons why this should have been so, prominent amongst them being asymmetries in the legal standing of the parties, the associated implication that international regulation of internal conflict might be construed as a form of interference in domestic jurisdiction, and, above all, the means of warfare associated with such conflicts. As a consequence of such various considerations, 'customary international law', as Roberts and Guelff express it, 'provided that the laws of war might become applicable to a non-international conflict through the doctrine of "recognition of belligerency" '.[11]

This has, however, begun to change in the post-war period. Even if regulation of irregular wars was not possible by the law of armed

---

[8] Ibid. 63.

[9] N. Fotion and G. Elfstrom, *Military Ethics* (Routledge & Kegan Paul, 1986), 213.

[10] W. V. O.'Brien, *The Conduct of Just and Limited War* (Praeger, 1981), 160; see also R. Higgins, 'International Law and Civil Conflict', in E. Luard (ed.), *The International Regulation of Civil Wars* (Thames & Hudson, 1972).

[11] A. Roberts and R. Guelff (eds.), *Documents on the Laws of War* (Clarendon Press, 1982), 12.

conflict, the introduction of various humanitarian practices is made possible by such devices as common Article 3 of the 1949 Geneva Conventions and the 1977 Geneva Protocol 11 goes even further in this direction.[12]

Why is it that irregular war seems to present distinctive problems for the implementation of codes of conduct? From a large field, three salient issues will be selected for scrutiny.

1. The most frequent complaint levelled against revolutionary or guerrilla warfare is that it is incompatible with a principle of discrimination in war. This is not necessarily in the sense that it employs indiscriminate terrorist tactics—although it may do so—but rather because the belligerents are 'concealed' amongst the civilians, a view perfectly captured in the image of the guerrilla as a fish swimming in the water of the people. What is wrong with such a practice? According to Phillips, it is a strategy which is doubly despicable both because it infringes discrimination while also presupposing the existence of such discrimination for its own success:

the success of guerrilla warfare presupposes *jus in bello*. . . . Because the guerrilla can, typically, count on his opponents to respect the immunity of civilians, his position is greatly reinforced. . . . The guerrilla is involved here in a double moral error: not only is he using civilians as mere means, but he is also relying upon the principle of noncombatant immunity while acting in such a way as to undermine it.[13]

As a moral theory of warfare, nurtured within a particular historical experience of war, this view has much to commend it. But it is opposed by another concept of war in terms of which this plea makes less sense. As we move on to the second perspective we will see that discrimination in war is very much a matter of political philosophy and that once again the philosophical protagonists operate within distinctive conceptual frameworks. Codes of practice cannot readily be nurtured in an alien conceptual environment.

2. This view can be stated bluntly in the assertion that regular war is the preferred instrument of the conventionally strong and to insist on respect for all the practices of conventional warfare is simply to express a political preference, wishing for the success of the status quo over revolutionary movements.[14]

---

[12] Ibid. 13.
[13] R. L. Phillips, *War and Justice* (University of Oklahoma Press, 1984), 98.
[14] This case is presented in Fotion and Elfstrom, *Military Ethics*, p. 213.

In a more subtle fashion, a similar case is made by analysing the political ideology of the revolutionary state, conveniently expressed by Walzer even though he does not accept its conclusions:

The guerrilla's self-image is not of a solitary fighter hiding among the people, but of a whole people mobilized for war. . . . If you want to fight against us, the guerrillas say, you are going to have to fight civilians, for you are not at war with an army but with a nation.[15]

If such a formulation causes disquiet, is it not best explained as a magnified form of that unease we feel when just warriors claim the right to dispense with the rules of war on account of the worth of their cause? What the case for guerrilla warfare does is to take the logic of this argument one stage further: it is the very nature of the struggle which defines the meaning of war and allows the conduct of the armed conflict to be drawn in its own image. Instead of claiming a monopoly on justice, the guerrilla seeks to make war his own and to dispense with those practices which no longer conform to the purpose.

3. Finally there are issues, less weighty in philosophical substance but no less significant in practical import, about guerrilla war's capacity to sustain the conventional laws of war. One example can be found in the area of POWs. Can guerrilla groups, logistically, take prisoners in the same way as conventional armies?

What this addresses is the more general issue of the authority, and political control, by means of which the laws of war are to be implemented. If the professionalization of armies, and associated codes of discipline, was to be an integral part of the development of the law of war, it is a moot point whether such guerrilla forces can be disciplined in the same way. The problem is not simply one of the acceptability of the code of war, but of its actual enactment. How are the codes to be enforced in irregular armies?[16] Not surprisingly, some commentators have insisted upon precisely such discipline and organization as the *sine qua non* of belligerence and of the applicability of the laws of war to insurgent movements.[17]

Instruments of War

Nowhere is the tension between military necessity and the rules of war more acute than in relation to the weapons that might be legitimately

[15] *Just and Unjust Wars*, p. 180.   [16] Ibid. 249.
[17] O'Brien, *The Conduct of Just and Limited War*, pp. 161–2.

employed in the waging of war. Given the widespread belief that available weapons will be used, the side which knowingly refrains from employing its full arsenal runs the risk of placing itself at a decisive military disadvantage. In more personal terms, the soldier who is not equipped with all available instruments of war is placed in mortal jeopardy. What right has the state to decree that soldiers shall risk their lives on its behalf and then deliberately intensify that risk by shackling the military efforts of its combatants and by denying them resort to all available means for getting on with the military task? Moreover, even if certain classes of chemical agents are not the preferred means of combat, is it not at least necessary to possess them as a deterrent to their use by the enemy?

The practicalities of implementing codes of combat are uniquely concerned with drawing lines, either on the basis of philosophically appealing categorizations or on the basis of significant, if somewhat arbitrary thresholds. This general procedural dilemma of the rules of war manifests itself in the particular context of the discussion of regulating the weapons of war.

Historically, the regulation of the instruments of warfare has followed two approaches.[18] The former is to outlaw broad, but unspecified, classes of weapons on the basis of general principles which some weapons infringe. The most important of such principles have been that instruments of war should not be treacherous, cause unnecessary suffering, be indiscriminate, or, more recently, that they should not have widespread or long-lasting ecological effects. The second approach is to ban specific weapon systems, such as the crossbow, gas, or other chemical agents. Presumably, however, to be plausible, such a specific ban must be coherent in terms of one or other of the general principles. The 1977 Geneva Protocols reaffirmed many of these general principles but this was not accompanied by agreement on the outlawing of specific weapons, despite a felt need in many quarters for such specific action to be taken.

On what basis should certain classes of weapons be prohibited? If it be in consequence of their indiscriminate effects, the objection to these weapons derives from a consideration of discrimination in warfare and can more appropriately be examined under that heading. Alternatively, it is on the basis of their cruelty and inhumaneness that some weapons come under scrutiny. Accordingly, the St Petersburg

---

[18] See A. Cassese, 'Means of Warfare: The Traditional and the New Law', in Cassese (ed.), *The New Humanitarian Law of Armed Conflict* (Editoriale Scientifica, 1979).

Declaration of 1868, renouncing the use of certain projectiles, set out the notion that weapons should not be employed which 'uselessly aggravate the sufferings of disabled men, or render their death inevitable'.[19]

This is, however, very dangerous philosophical territory into which to stray. The temptation to draw arbitrary lines is strong given the manifest difficulties in differentiating between marginal increments of pain and suffering, and measuring one kind of mutilation against another. On what principle do we determine that a tumbling bullet, capable of inflicting massive intestinal wounding, is a humane and legal weapon, whereas certain forms of chemical weaponry, producing instantaneous death, are to be proscribed?

Faced with this complexity, the law of armed force has pursued a strategy of satisficing by drawing upon a measure of philosophical principle, combined with elements of common practice and tradition, and measured against the likelihood of acceptance. As a generalization it can be said that the only weapons in the history of war whose deployment, or use in war, has been successfully regulated by international agreements have been those weapons which were of dubious military value and the use of which the belligerents have been prepared to forgo in any case. Where states do see advantage in employing illegal weapons they may do so while denying that they are, as the use of chemical agents in the Gulf War has revealed.

An assessment of the acceptability or otherwise of specific weapon systems may vary depending on whether a just war, or limited war, perspective is applied. The just war demands of proportionality and its insistence that the war be fought with a right intention can both be read to preclude excessive cruelty. It is more difficult to see how humaneness is incorporated in a limited war doctrine, except as a function of avoiding militarily unnecessary acts of war and forestalling the spiral of reprisals that might flow from them. But if an act of cruelty would serve to limit a war, it more obviously falls under the ban of just war theory than that of limited war.

Principles of Discrimination

In the eyes of some, respect for a principle of discrimination is the cardinal requirement of any doctrine of just or limited war. According

---

[19] Roberts and Guelff, *Documents on the Laws of War*, p. 31.

to the argument of Michael Walzer, war is the realm of moral choice and the most important moral choices that we need to make concern the nature of the targets against which the war will be prosecuted.[20] It is respect for such a principle that marks war off from other forms of killing. This is to be contrasted with the argument of Hobbes which is equally insistent that the enemy is to be classed as one and that no basis exists for distinguishing between members of the enemy state in such a way that would affect our conduct of the war. It is part of our conception of war, according to Hobbes, that 'the sword judgeth not, nor doth the victor make distinction of nocent, and innocent'. If the Hobbesian position is not accepted, how might a principle of discrimination in war be articulated?

Modes of Discrimination

Political discrimination The basis of this principle is the requirement that a distinction be made between the political leaders responsible for engaging the state in war and the remainder of the (politically) innocent population. Thus Thomas More, writing in his *Utopia*, claimed that 'the common folk do not go to war of their own accord but are driven to it by the madness of kings', and in saying this he was expressing what was to become the stock liberal position on war, namely that it was a product of corrupt 'government' and not the expression of popular will. The whole strand of the Wilsonian New Diplomacy was predicated on eliminating war by bridging the gulf which separated government from people in the conduct of foreign affairs. To this extent, Wilsonianism itself accepted that discrimination could be made between (unrepresentative) rulers and their people.

The problem with this particular form of discrimination is, of course, that leaders tend to be the least accessible members of the enemy state for purpose of directing military activities against them. Frequently the capital city, or seat of government, is the last enemy target to make itself available to military operations and was certainly so before the development of modern airpower. There is therefore a logical gap in this theory because, as far as the traditional practice of warfare is concerned, the principle tells us nothing about what may be done to the populace in general in the effort to get at its leaders.

It may none the less be argued that this problem has been reduced in the missile age. Indeed, in recent years, US strategic targeting

---

[20] *Just and Unjust Wars*, p. 42.

doctrine, at least in its publicized form, has apparently moved in the direction of giving increased emphasis to the destruction of the Soviet political and military command structure in the event of nuclear hostilities. The basis of the reasoning for this would seem to be that since, given the nature of the Soviet political system, what the leadership values most is its own continuation, rather than the lives of its citizenry, it makes sense to threaten to destroy the leadership. However, if missiles allow war to be waged directly against enemy leaders in their war bunkers, without the necessity of first bursting through their lines of territorial defence, the use of nuclear weapons presents further potential problems for the effort to discriminate between leaders and people, as will be discussed in the next chapter. In any case, it is a moot point whether this change in targeting doctrine is based on a principle that this is how war ought to be conducted properly or whether it merely reflects the instrumental calculation that presenting a threat of this kind is likely to be the most effective deterrent.

Institutional discrimination This principle is based on the contention that there is a fundamental philosophic distinction between the institution of the state and the society or people who are contained within it. Its clearest formulation was by Rousseau who argued vehemently that war is a relation between state and state and not between people *per se*:

War then is a relation, not between man and man, but between State and State ... each State can have for enemies only other States, and not men; for between things disparate in nature there can be no real relation.

Rousseau's is a bold and imaginative philosophic conception. What is not quite so clear is how the principle is to be observed in practice and, even allowing that it was materially possible to do so in the conditions of eighteenth-century warfare, it is desperately unclear how it is to be made viable in the conditions of the twentieth century.

This is so both for practical reasons, flowing from the manner in which recent wars have been conducted, but also for other reasons, having a basis in political philosophy, which call Rousseau's distinction into increasing question. Best summarizes them in the following passage:

The distinction between the combatants of the armed forces and the 'peaceful' rest of the population, conceptually indispensable for a law of war that is to mean anything, has always found itself jostling for living-room with the more aggressive sorts of democratic, patriotic and nationalist rhetoric, which imply if they do not actually affirm that there really is no significant split between the combatant at the front and his family and neighbours at home. . .[21]

The actual practice of warfare has likewise, in the wake of developments in military technology, moved in the very opposite direction from that which Rousseau's principle would entail. While Rousseau believed that 'sometimes it is possible to kill the state without killing a single one of its members', military practitioners have found that killing its members is by far the most convenient way of setting about the business of destroying the enemy state, as the British naval staff candidly asserted in 1921:

Nothing can be clearer than the fact that modern war resolves itself into an attempt to throttle the national life. Waged by the whole power of the nation, its ultimate object is to bring pressure on the mass of the enemy people, distressing them by every possible means so as to compel the enemy's government to submit to terms.[22]

Such reasoning reached its culminations in the strategic bombing policy of the Second World War, directed against the morale of the population, and eventually in the atomic bombings of Hiroshima and Nagasaki.

Manifestly the fact that a principle is ignored in practice is no reason whatsoever for questioning its theoretical or philosophical force. Presumably the laws of war are intended to shape the practice of waging war and not simply to reflect what is actually done. It should also be emphasized, therefore, that Rousseau's principle is challenged, not by the twentieth-century's experience with 'total war' alone but also by the tendencies in the political theory of the state itself which, in many respects, has served to erode the sharpness of the distinction between state and society on which Rousseau's argument is based. This is certainly not to suggest that state and society are coextensive categories but simply to point out that the business of separating them out, as a matter of practice, is that much more difficult after the past half-century's experience with the increasing role of the state in a variety of social functions.

Moral discrimination This principle is based on the idea that it is possible to discriminate between those whose actions are morally

[21] G. Best, *Humanity in Warfare* (Weidenfeld & Nicolson, 1980), 222.
[22] Quoted in B. Bond, *War and Society in Europe, 1870–1970* (Fontana, 1984), 144.

culpable, either in bringing about the state of war or in prosecuting it in an improper way. If this distinction can be made, it follows, according to this argument, that the pain and suffering of war should be directed against those who are morally responsible. It therefore has intellectual roots in the just war conception of war as an instrument of punishment of the evil-doer and, since this is the point of a just war, it is fitting that its anger should be directed against the evil-doers themselves.

Little more need be said to indicate that, as a practical system for controlling the conduct of war, there are difficulties with this notion. Moral guilt is not, for the most part, an externally recognizable characteristic. How then is it to be operationalized as a way of confining hostilities to a distinct category of people? While it may have considerable appeal as a moral injunction, this principle seems to fail all tests of saliency without which the implementation of any type of discrimination in warfare becomes virtually impossible.

Given this shortcoming, its area of application seems to have been narrowed, in practice, to the determination of guilt in the aftermath of war rather than in the course of it and has become transmuted as one of the possible bases for assessing war crimes. This is in recognition of the fact that any attempt to assess responsibility on the enemy side while blinded by the fog of war is futile and best left until visibility has improved in its aftermath. This is the more so when responsibility is being defined not in formal, but in strictly moral, terms.

Military discrimination This principle expresses itself in the classical distinction between combatants and non-combatants and employs this distinction to advance a claim for non-combatant immunity. Its general tendency is to confine actual hostilities to the combatants themselves, in this case defined as those who are engaged in the business of bearing arms.

Traditionally, the great virtue of this distinction was its salience. For instance, it can be associated with the development of distinct military uniforms. Uniforms perform a dual function both by way of a liability and a protection. They are a liability inasmuch as they advertise the wearer's availability as a legitimate target of war. They also serve as a protection by guaranteeing the wearer certain rights, encoded in the laws of war, while he is out of the combat, most notably rights to reasonable treatment as a prisoner of war and rights to medical attention if wounded. These rights were traditionally not guaranteed to the irregular fighter whose status was not marked by the wearing of a uniform and who was, to that extent, exploiting the status of civilian. In doing so, the irregular fighter forfeits the rights of a warrior.

Again, however, the technical state of the military art has cast doubt upon such a straightforward distinction. The wearing of a uniform on a field of battle is not sufficient to define the category of combatants. The mobilization of an entire population behind the war effort, and the imagery of the 'home front', makes the traditional conception seem superficial and lacking in philosophic substance. For example, was a cypher clerk working at Bletchley, engaged in cracking enemy codes, to be classed as a combatant or not? Certainly he did not bear arms: equally, his efforts and activities led directly to the infliction of substantial casualties upon the enemy and, to that extent, was probably more dangerous to the enemy than any front-line soldier.

The search for modes of discrimination in war is a philosophic debate about how, and where, to draw lines between various categories. What is deficient with the principle of non-combatancy is that the drawing of that particular line immediately requires the search for another set of more pragmatic contours.

Principles and Practice of Discrimination

Do sound practices of discrimination in warfare need to be based upon equally sound philosophic principles? Or is it enough that a practice has some customary basis even if devoid of compelling philosophic content? This general question has been posed in the specific context of discussions about the nature of the principle of non-combatant immunity. Is this a universal norm or simply a convenient, and to some extent accidental, practice? Hartigan has pressed the latter position.[23] For him, the practice of respecting the rights of non-combatants developed, for historical reasons, in a period of warfare when it was convenient and sensible to make some effort to wage war in this way. There was physically some separation between civilians and the field of battle and the nature, say, of eighteenth-century 'limited wars' encouraged the practice for other prudential and economic reasons. It should not, however, be mistaken for a moral axiom. In Hartigan's words, 'there is certainly room for legitimate doubt as to whether the

    [23] R. S. Hartigan, 'Noncombatant Immunity: Its Scope and Development', *Continuum*, Aug. 1965, 214–18.

norm of civilian immunity enjoys the status of an absolute moral imperative.'

This suggests that the principle is to be regarded as one of utility rather than one of morality. It makes sense to observe it where possible but the fundamental nature of war would not be damaged if the principle is infringed. But as previously explained, is the matter so simple? If it is morally desirable to have rules of discrimination, and if it is equally essential that they be practicable and enforceable rules, is the moral force of the principle diminished by being *merely* an extant practice and by lacking a degree of internal moral coherence? May not its moral standing derive from the absence of any other distinction which we might hope to implement? If we are to moderate war and this goal is seen to possess moral value in and of itself, then we need enforceable rules even if they, in and of themselves, lack moral coherence.

It might also be objected that, by framing the principle of non-combatant immunity in such a way that we focus upon the rights of non-combatants, this may inadvertently distract our attention from the equally important rights of the combatants. In saying that combatants are legitimate targets of warfare, do we not unwittingly undercut any rationale for limiting what may be done to these targets? In any case, which is the greater evil—the humane killing of non-combatants or the burning of combatants to death by flame-throwers? Such a moral calculus is grisly but it should not be ignored and the principle of non-combatant immunity, if advanced as the sole principle of moral warfare, carries the danger of leading us to ignore it. Happily the two issues are by no means as incompatible as this argument suggests. The fact that a combatant, as a legitimate target of war, loses some rights, including that not to be killed, does not mean that he loses all rights. He may retain other humanitarian rights, including rights to dignity, even in the moment of his death. Although some may find even this a paradoxical idea—that the soldier should lose his right to life (arguably the most important right of all) and yet retain some rights over the manner of his death (possibly a right of the second order).

Whichever principle of discrimination we seek to apply, there is a further fundamental choice to be made: where does the onus of proving the validity of the claim to immunity lie? The point may be made by way of analogy with the legal system which can proceed on the assumption that the accused is innocent until proven guilty or, alternatively, may work on the assumption that the accused is guilty until proven innocent. In accordance with British practice the onus rests with the Crown to establish the guilt of the accused and, if this case fails, the accused remains innocent. This is to say that his status remains unchanged. In other legal systems the opposite procedure holds and the onus rests with the accused to demonstrate his innocence. If he succeeds, he changes his status from guilty to innocent.

There is some parallel here with the procedures which might be adopted for establishing, or denying, the validity of claims to immunity in war. On the analogy of remaining innocent until proven otherwise, it might be inferred that all participants in war are deemed to have a right to immunity and the onus rests upon the enemy state to deny that claim or to demonstrate why, in the circumstances, the claim should be overridden by appeal to some higher principle. Alternatively, it might be inferred that all persons are legitimate targets of warfare and the onus rests upon each and every individual to make a case why he or she should be deemed 'innocent' and be regarded as immune.

There are fundamental differences entailed by these distinct procedures. The one takes as its point of philosophical departure the assumption of immunity for all unless *we* can deny their claim to it in specific instances. The other assumes that there is immunity for none unless *they*, individually, can establish some claim to it. In the former, targets are progressively added; in the latter, they must be progressively eliminated.

The practical importance of this difference is that, in the absence of evidence which would allow a case to be made either way, the one assumes the entire population to be innocent, the other that the entire population is guilty. Are we not here back once again with a distinction which is similar to the one advanced by Paskins that it is the business of just war to determine, not how far its reach may be limited, but how far its reach may legitimately be extended?

To the extent that these parallels are exact, it might be concluded that limited war theory tends to favour the procedure of assuming the guilt of all, while working to eliminate certain categories from the list, whereas just war theory assumes the innocence of all, except for those categories than can legitimately be added to it. This is the crux of the moral difference and touches further upon the issue of the morality of having any rule versus the morality of the content of specific rules. It surely requires a less persuasive moral case for sparing an individual from his death than it does for condemning him to it? In other words, it is easier to find a moral purpose in the development of purely utilitarian categories of immunity than it is in finding any moral sanction for using utilitarian grounds to deny valid claims to immunity. It matters less in the former case if we act on coherent moral principles because moral benefit will derive from our actions *whatever* we do. Applied to the business of denying claims to immunity, such a procedure is morally execrable.

The question of where the moral onus is to be located in warfare manifests itself in other practical ways. For instance, there is widespread condemnation of guerrilla warfare because its nature is such as to render impossible the practice of discrimination. Since guerrilla warfare, by definition, collapses the distinction between combatant and non-combatant by the way in which the guerrilla may seek to hide among the non-combatants, the war cannot be prosecuted by observing this distinction. What precisely is the nature of this objection? If discrimination is a necessary part of war as a legitimate procedure, then what is morally objectionable is not what the guerrilla does to the enemy but what he does not do: he does not offer his enemy the moral choice of prosecuting the war against combatants and not against non-combatants. To the extent that this argument is compelling, it urges that our moral responsibility goes beyond the manner in which we wage our own war. It obliges us also to take such steps as are necessary to allow the enemy the opportunity of justly conducting his own.

This same issue recurs in some of the recent debate about nuclear strategy. To the extent that discussion about nuclear war, not to say planning for such war, has proceeded on the basis of making some distinction, however tenuous, between countervalue and counterforce targets, it could be argued that it is as equally immoral to collocate military installations near major centres of civilian population (thus denying the enemy his right to a just war) as it is to prosecute a war with no effort whatsoever to discriminate between different kinds of targets. In other words, the debate about what kind of conduct is becoming, or unbecoming, in warfare is not simply about what should be done: it is also a debate about where the moral onus rests for providing the requisite conditions in which legitimate warfare can be pursued.

Finally, two points are in order in connection with the political philosophy of the democratic state. Is it not possible that there is a tension between the general tendencies of democratic development and the attempt to establish immunities in war? While it was tenable to posit a clear line of demarcation between the autocratic monarch and his people, is not this line called into question by the theory, if not by the actual practice, of the democratic state? If war is regarded as an expression of popular will, represented through authoritative political procedures, is there enough philosophical space left between leaders and followers, between the guilty and the innocent, or between state and society, to continue to accommodate a theory of immunity? Has not the tendency of democratic thought been to erode any meaningful distinction between most of these categories? Are not wars more genuinely struggles between peoples, *pace* Rousseau, if they are fought between democratic states, and on what basis of right does the individual opt in to the policies of the state but opt out from the brunt of their military consequences?

If there is any force to this particular argument, it has a second and related corollary in terms of discrimination in war. If there is a sense in which a citizen's availability as a legitimate target of war is related to the freedom of the political institutions of the state, then there may well be a case for saying that civilians in a democratic state offer more appropriate targets of warfare than do citizens in undemocratic ones because the war is more properly their own: they share responsibility for its initiation and, on some theories of people's war, share equally in its prosecution. Rather than argue for a constant set of principles of discrimination, applicable in all wars, do we not then also require a concept of asymmetric justice in war whereby what can legitimately be done is at least partially dependent upon the nature of the state against which the action is being taken? If, for instance, we believe, as some of our political rhetoric suggests, that Soviet citizens are little more than the prisoners of their own political leadership, is not there a duty on Western countries, in developing their war plans, to try to respect the right of 'innocence' of the Soviet citizenry? In the meantime, no similar onus rests upon the shoulders of the Soviet leadership. They are liberated in their conduct of a war by the palpable claims of our own democratic institutions.

Such a conclusion may obviously seem unpalatable on a number of grounds. What the argument does is to conflate, in assessing the guilt of Western citizens and the innocence of their Soviet counterparts, various forms of discrimination. For instance, if non-combatancy is taken as the touchstone for the waging of war, it is in no way apparent that the participation of Western citizens in democratic political decision-making should make them any more legitimate as targets of military activity. Perhaps, in any case, the guilt of Soviet citizenry resides in their tolerance of their own regime and in the fact that they do not strive more assiduously to liberate themselves from it, in much the same way as some condoned the bombing of German cities because of the tacit guilt all Germans shared by their non-resistance to the Nazi tyranny.

The trouble with such notions, in turn, is that it is but a short step from them to espousing the original 'holy war' idea that it would be a legitimate function of a targeting policy to destroy the opponent's regime or to stimulate a war of liberation (for instance by targeting the structures by means of which the USSR's ethnic populations are held in subservience to their Russian masters). There is then a dangerously thin line between discriminating between leadership and populace for reasons of assessing responsibility (just as with superior orders in a formalized military command structure) and discriminating between them as part of a policy of external intervention and liberation.

Enough has been said to indicate the flavour of the dilemmas posed and to suggest that the practicalities of implementing codes of war are determined by a range of competing and countervailing considerations. These extend across the broad spectrum of issues raised in contemporary political theory about the nature of the state and the relationship of citizens to this collective vessel. The problems of implementation are doubly difficult in an age of warfare in which it is anticipated that intra-war communication and bargaining would have a high profile and upon the successful outcome of which any final termination of a war might depend. The reason for this is that in a relationship of such tacit communication, it is necessary not only to recognize the philosophical basis of one's own actions but to convey this information to the enemy. A legitimate Western attempt to spare Soviet citizens might be interpreted as a more hostile act in the eyes of the Soviet leadership and it is to be doubted that acts of war have the necessary clarity for conducting a seminar on the niceties of democratic theory in such an inauspicious setting as that of an ongoing nuclear war.

5

CONTEMPORARY NUCLEAR STRATEGY AND WAGING WAR

THE prospect of nuclear war, and the place of such a prospect in nuclear strategies of deterrence, brings together all the fundamental issues hitherto considered. It focuses our attention upon principles of discrimination and proportionality in warfare. It asks the question whether the justice of war can reside in the very nature of the weapons by means of which it is fought. Beyond this it probes the relevance of traditional ideas about limited war and tests their applicability, both in practical and in moral terms, to the conditions of nuclear war. More acutely than ever before it presses us to consider the precise nature of our intentions and whether, by making preparations for war in order to avoid it, we are really intending to fight such a war or not. Most essential of all, it invites us to think carefully about the goals and purposes of war and to determine how far we may legitimately go, and how far we are actually prepared to go, in attaining these objectives. In short, it is impossible to make any intelligible statement about the conduct of a nuclear war without some attempt to relate it to a wider conception of war itself. To do so may not be a sufficient, but it is at least a necessary, condition for resolving our philosophical dilemmas.

There has been a widespread, though not universal, presumption that such is the destructive capacity of nuclear weapons that this must lead to a dramatic narrowing, if not complete elimination, of those just causes which would permit the undertaking of war in which those weapons would be used. In turn, such assessments have derived from the perception that nuclear weapons are 'unique' in the history of warfare, a view reiterated in Nye's recent injunction 'never treat nuclear weapons as normal weapons'.[1] The solemn priority that Soviet and US negotiators assign to the arms control of such weapons, as well as President Reagan's self-proclaimed goal of making them 'impotent and obsolete', each in its own way tends to reinforce this image of

[1] J. Nye, *Nuclear Ethics* (Free Press, 1986), 105.

distinctiveness for nuclear weapons within contemporary military arsenals.

Not all writing on nuclear weaponry uncritically accepts such an initial assumption. Much of the writing on nuclear strategy and warfare over the past two decades has operated on an assumption that nuclear weapons can be integrated into traditional military categories, or, in the words of one critic, has recommended 'not that we alter our behaviour in the wake of knowledge of nuclear technology, but rather that nuclear weapons and strategy be accommodated to our behaviour'.[2] Others have insisted that to talk of nuclear weapons as a homogeneous category is to commit a palpable error as there are 'important and morally significant differences between the various kinds of nuclear weapons and their possible uses',[3] albeit the significance of these differences remaining contingent upon the escalation that may, or may not, take place. In this chapter, this general issue can be investigated by considering the subject of nuclear targeting, other salient characteristics of nuclear war-fighting doctrines, and problems associated with the attempt to make judgements about war purely on the basis of individual weapon systems, such as nuclear arms.

Strategy and Targeting

Nuclear strategy has been subjected to the widespread criticism that it is inconsistent with any principle of discrimination in war. This was deemed to be so for two reasons. First, it was a consequence of the physical nature of the weapons and the scale of destruction that would be wrought by them. Secondly, however, it was charged that the violation of discrimination in connection with nuclear weapons was a product of misguided strategic choices, rather than an inevitable fact of nature. If the latter was so, more discriminating nuclear strategies might be adopted.

The issue had been crystallized by the debate about the SALT 1 Treaty in 1972 and by the seeming formal acceptance of the doctrine of mutual assured destruction. The treaty appeared to sanction the hostage relationship whereby the populations of both superpowers would be held in permanent vulnerability as hostages for the good behaviour of their respective governments. The fact that, by the ABM

---

[2] P. K. Lawrence, 'Nuclear Strategy and Political Theory: A Critical Assessment', *Review of International Studies*, Apr. 1985, 115.

[3] D. Fisher, *Morality and the Bomb* (St Martin's Press, 1985), 53.

Agreement, the superpowers were denying themselves the opportunity of developing defensive systems against missile attack in the future, lent force to this particular interpretation.

The logic of the posture was that war would best be deterred by the absolute nature of the threat which it entailed: if war should break out, it would be a war of mutual destruction with casualties running into such very high millions as to render entirely inappropriate any pretence that the war was being conducted in conformity with principles of discrimination. On the contrary, the only targets likely to be spared destruction would be hardened military targets whilst the hapless unarmed, and unprotected, civilians would bear the full ferocity of the attack. Albert Wohlstetter was to comment incisively upon this cruel paradox that 'not even Genghis Khan tried to avoid military targets and to concentrate *only* on killing civilians'.[4] The intention was not, of course, directly to kill millions of civilians. But the harsh reality underlying the deterrent intention could not be ignored. Fred Ikle had drawn attention to this poignant dilemma in his observation that 'it is a tragic paradox of our age that the highly humane objective of preventing nuclear war is served by a military doctrine and engines of destruction whose very purpose is to inflict genocide'.[5]

In other words, what seemed to be wrong with this strategy was that it placed all its eggs in the deterrent basket and made little or no provision for the eventuality of the threat having to be executed. Once the business of fighting a nuclear war in accordance with the prescriptions of assured destruction was looked at closely, it seemed less than palatable. Could such a war conform with any standard of moral propriety or political instrumentality?

The balance of the argument suggested not. This did not mean that there was a universal conversion of nuclear strategists away from the concept of assured destruction but the sentiment was sufficiently influential to prompt the search for an alternative strategy, the major virtue of which was to be its limitability, not least in the area of discrimination.

Since the early 1970s successive US administrations have devoted attention to reformulations of their declaratory nuclear strategies. There have been differing emphases over the period but also some

---

[4] A. Wohlstetter, 'Threats and Promises of Peace', *Orbis*, Winter 1974, 1127.
[5] F. Ikle, 'Can Nuclear Deterrence Last Out the Century?', in R. Pranger and R. Labrie (eds.), *Nuclear Strategy and National Security* (American Enterprise Institute, 1977), 130.

themes in common. Their collective hallmark has been the search for a strategy that would best reconcile the requirements of deterrence with the requirements of fighting a war should the deterrent fail. And since whatever virtues assured destruction of populations might have possessed lay in the field of deterrence, and not execution, this has amounted to the presentation of a new strategy whereby the US must possess the physical capability to prosecute a limited, controlled, and possibly protracted nuclear war. The emphasis throughout has been on developing flexibility and a menu of limited options. For the most part this strategic revisionism has manifested itself in the designing of new sets of nuclear targets.

To some extent such a doctrinal choice has only become possible in the past decade as major developments in missile technology, particularly with regard to accuracy, have become available. There would have been little practical point in proposing selective nuclear strategies designed to discriminate between military and population/ industrial targets at a time when the weapons themselves were incapable of achieving anything like the degree of accuracy which such selectiveness would presuppose. Accordingly, vocal proponents of the new strategies of discrimination, such as Albert Wohlstetter,[6] have spoken of a new revolution in accuracy, to add to the original nuclear revolution in terms of explosive power.

Originally the suggestion seemed to be that a nuclear war might be limited both as regards targets and as regards the tempo of hostilities. Major attention would be devoted to the destruction of the enemy's own military infrastructure and, according to the scenarios of the mid-1970s, the release of nuclear weapons would not be in a single spasm but would be highly controlled and modulated to the requirements of political bargaining. Latterly, US officials have spoken more in terms of targeting the political and military command centres of the USSR as a priority objective. Even the British Government, which by tradition says little at all about its own targeting policy, had adopted the formula of seeking to destroy the organs of Soviet state power, which at least implies an intention to discriminate between the Soviet state and its people more generally.

To adopt a principle of non-population targeting is not, of course, to make this a reality. To the extent that post-1945 theories of nuclear deterrence have represented a carry-over from wartime strategies of

⁶ 'Bishops, Statesmen, and other Strategists on the Bombing of Innocents', *Commentary*, June 1983.

area bombing, they are a manifestation of technical constraints rather than of free strategic choices. It is certainly the case that the Second World War strategic bombing was dictated by technical impediments to precision bombing and only later was this policy provided with the rationale of destroying morale and the will to continue with the war. The direction of post-war nuclear strategies, bounded by the characteristics of first-generation atomic bombs as city-busters, was determined by technical factors and did not embody any clear strategic preference for population targeting. As early as January 1946 Britain's Chiefs of Staff had been advised by an Atomic Weapons Subcommittee that, for technical reasons of cost-effectiveness, 'only important targets would be worth a bomb' and, therefore, it should be expected that attacks on large cities would 'become the most profitable type of war'.[7]

None the less, those same technical parameters continue to call into question those strategies which would move targeting away from concentration on civilian targets. The choice between the Soviet leadership and the Soviet people is more clear-cut in theory than it is in practice. As a recent authoritative study has argued '[i]t would be difficult to destroy any extensive segment of the Soviet leadership without causing large numbers of civilian casualties'.[8]

More generally, it has been claimed that the Soviet population *per se* is not a listed target in US nuclear war plans. While this may well be the case, it is by no means clear how significant this fact would turn out to be in the event of war. Given the collocation of target systems with centres of population, civilian populations would suffer high casualties regardless of the specific targeting doctrine. As Jeffrey Richelson has claimed, a prohibition on population targeting would tell us more about targeting mechanisms than about actual effects.[9]

Additionally, as with other aspects of nuclear strategy, there might even be costs in trying to abandon counter-city postures and any assessment of such trade-offs is difficult to calculate. If the side-effects of counterforce doctrines, adopted to allow more targeting precision and hence to minimize casualties in the event of use, are the greater strategic instability associated with counterforce policies and a proclivity to pre-emption, it is not at all clear that the cause of making civilian lives more secure has been advanced by their adoption.

---

[7] Public Record Office, CAB 82/26, DCOS(AWC) (46)1, 30 Jan. 1946.

[8] D. Ball in Ball and J. Richelson (eds.), *Strategic Nuclear Targeting* (Cornell University Press, 1986), 23.

[9] Ibid. 28.

Most of the debate about the relative virtues of the old and the new strategies was conducted in technical strategic terms which made little direct reference to the wider issues entailed by philosophical discussion of war. Strategic revisionism was supported for sundry reasons which included the increased deterrent effect of issuing a more limited and, thereby, more credible threat, for its merit in thinking beyond deterrence alone into what to do in the realm of a war actually initiated, and because a de-emphasis of deterrence-only strategy would put the US's strategy on a doctrinal par with that of the USSR which had never articulated the same formal and abstract theory of deterrence as had developed in the United States. Above all, it was predicated upon a rejection of assured destruction, both because of its incredibility as a deterrent to calculated war and because of its potentially suicidal results in the event of inadvertent or accidental war.[10]

As against this, the critics lamented the increased risk of nuclear war that might inadvertently be the consequence of a seemingly more innocuous form of nuclear hostility, and demurred from the principle that doctrine should unthinkingly follow wherever military technology should lead. They also admonished that the new strategy was but a prescription for a new arms race and for a dangerous crisis instability (since counter-military targeting favoured the side which fired first) especially because, as if in poetic justice, the Soviet leadership stoutly resisted any conception of limited nuclear war.

A prominent theme of commentaries, in this regard, was that strategies of limited nuclear war, designed to strengthen deterrence, would most likely end up by undermining it as the assured destruction relationship came to be destabilized. 'The absurd struggle to improve the ability to wage "the war that cannot be fought" ', two high-profile participants remarked, 'has shaken confidence in the ability to avert that war.'[11] Others reached similar conclusions but for different reasons, arguing that assured destruction was inescapable in any event as the 'nuclear revolution cannot be undone' and, in consequence, the United States 'is deterred by the fact that its cities are vulnerable, not by the fact that the Russians have some supposed military advantage'.[12]

[10] Nye, *Nuclear Ethics*, p. 110.
[11] R. S. McNamara and H. Bethe, 'Reducing the Risk of Nuclear War', *Atlantic Monthly*, July 1985, 45.
[12] R. Jervis, 'Why Nuclear Superiority Doesn't Matter', *Political Science Quarterly*, Winter 1979–80, 631.

While most of the debate was of this order, and the official debate certainly so, there none the less did occasionally shine through an alternative perspective. I have referred to this elsewhere[13] as the school of 'moral counterforce' writers who have urged, on explicitly moral or just war grounds, that nuclear strategy be made, as far as is possible, to conform to the requirements of discrimination. While Walzer has been persuaded that nuclear weapons have exploded the doctrine of just war, others have tenaciously maintained that the principles of that doctrine continue as a basis of action even while conceding that there may be problems in conforming to them in practice. Paul Ramsey was an early exponent of this view and urged the adoption of a counterforce nuclear posture because, as he put it, 'I cannot see how a city-targeting strategy can possibly be reconciled with principles of just employment of armed force.'[14] In Britain, the argument was pressed by proponents of graduated deterrence that massive retaliation was unacceptable on political and military, but also on ethical, grounds: 'Morally, we should not cause or threaten to cause more destruction than is necessary. . . . Massive retaliation hardly passes this test, nor indeed does it square with the moral standards we professed to uphold at the Nuremberg trials.'[15] More recently, James Johnson has shared this concern in urging that 'reasonable options exist to take us away from counter-population targeting toward a more morally defensible policy'.[16]

How morally acceptable some of these counterforce alternatives might be has, in turn, been submitted to close scrutiny. Ramsey, in particular, was aware that a purely counterforce deterrent might not be as effective against the enemy and, therefore, while decrying city-targeting, suggested that the knowledge of the collateral civilian deaths that would be incurred even in a counterforce-only attack would itself add to the effectiveness of the deterrent. He seems, to his critics, to have wanted 'incidentally' to benefit from an effect which he deplored but also accepted as 'inevitable'. In these circumstances could double effect be appealed to with a clear conscience? His critics think not:

Since we know that counterforce strategy will deter aggression only if the threat to the civilian population is very great, and since we intend these threats

---

[13] *Limited Nuclear War* (Princeton University Press, 1982), 173–6.
[14] See e.g. Ramsey's *The Limits of Nuclear War* (Council on Religion and International Affairs, 1963) and *The Just War* (Charles Scribner's Sons, 1968).
[15] A. Buzzard, 'The H-Bomb, Massive Retaliation, and Graduated Deterrence', *International Affairs*, Apr. 1956, 154.
[16] J. T. Johnson, *Can Modern War Be Just?* (Yale University Press, 1984), 148.

to deter, it follows that we have formed the conditional intention to bring about their deaths after all.[17]

Partly to compensate for this perceived defect, and partly to overcome the objection that the moral virtue of counterforce strategy is more than offset by the strategic instability it induces, Nye has urged a counter-military, rather than a counter-silo, strategy. Nuclear forces, under this rubric, would be targeted against advancing Soviet armies, as well as at logistics and key military routes. This would require a strategic force larger than that needed for assured destruction, but smaller than that needed for a counter-silo strike.[18]

In the terms of this debate, which is at once narrower and broader than the technical one, it would seem that some moral trade-off may need to be made. The force of the argument is that such a posture is in greater conformity with traditional ideas about warfare if ever the threat of nuclear war had to be implemented. And yet somehow the greater credibility of such a threat seems dangerously close to a heightened risk of war. In the great accounting, how do we compute the moral value of lives saved by threat of immoral war as against lives that may just conceivably be lost by a war which succeeds in conducting itself to moral precept but fails momentously in not preventing its own outbreak?

Beyond this, it may also be the case that targeting strategies have to bear the burden not only of war prevention and limitation but also of war termination. Given the nature of nuclear war, whatever other objective is pursued by means of it, a paramount goal must be to bring it to an end as rapidly as possible. Just as some deterrent strategies do not make for sensible military practice if executed, likewise not all deterrent strategies, either counter-population or counter-leadership targeting, necessarily make for plausible war termination. Arguably this has been the most neglected aspect of the nuclear strategy debate. As George Quester has recently expressed the case,

Advocates of 'winning' a nuclear war have placed great stress on the need for imaginative thinking in contemplating the target base of the Soviet Union. Advocates of terminating such a war should be encouraged to join in this exercise with similar imagination and zest. Preventing the outbreak of such a war in the first place still commands the highest priority, but preventing its

---

[17] R. P. Churchill, 'Nuclear Arms as a Philosophical and Moral Issue', *Annals* (American Academy of Political and Social Sciences), Sept. 1983, 54.
[18] *Nuclear Ethics*, p. 112.

continuation into a second day or a second week might also merit a peace prize or two.[19]

There is also the objection that discussion of discrimination in a nuclear context is but a form of verbal casuistry. We may employ the same words, but they must have a radically altered meaning. In effect, this is to say that we cannot tame nuclear weapons by mere dint of changing our nuclear doctrines. What makes nuclear weapons resistant to any meaningful discrimination is not the imperfections, or otherwise, of the doctrines which have been developed for their use, but is rather a part of their intrinsic nature. To consider this matter we must turn to other issues raised in the discussion of fighting a nuclear war and then to the specific matter of the nature of individual weapons systems and their place in conceptions of just or limited war.

Issues in Nuclear War-Fighting

The attempt to develop discriminating nuclear targeting strategies has been part of a larger endeavour to accommodate nuclear weapons to operational military tasks. It has been driven, above all, by the felt need to relate deterrence more closely to military capabilities but, in doing so, has created anxiety that national strategies are beginning to be based on the intention to fight a nuclear war. 'The dream of fighting nuclear war rationally, even a defensive war within moral limits reconcilable with human values,' it has been written, 'is part of a vast nightmare into which we are drifting.'[20] That the professional literature on nuclear strategy, during the past decade, has become more singularly preoccupied with aspects of *waging war*, there can be little doubt: whether this betrays a more prevalent acceptance of the risks of undertaking such war is an open question but the perception that some such policy change was occurring was intimately connected with the widespread nuclear phobia of the late 1970s and early 1980s.

Whether such a strategic emphasis, in order to avert all wars in general or nuclear war in particular, is an acceptable mode of strategic operation is a question best deferred until the survey of deterrence presented below. However, this perspective has focused professional, as well as public, attention upon a number of issues involved in creating the requisite strategic capabilities for the conduct of nuclear war and these can be briefly outlined at this point.

---

[19] In Ball and Richelson, *Strategic Nuclear Targeting*, p. 305.
[20] R. F. Rizzo, 'Nuclear War: The Moral Dilemma', *Cross Currents*, Spring 1982, 73.

The notion that nuclear strategy must think through the detailed operational sequences of nuclear attacks, rather than simply relying on absolute levels of destruction, has long been associated with the view of the balance of terror as 'delicate' and survived the orthodoxy of assured destruction articulated in the McNamara years. It emerged powerfully in the later 1970s, not least as an attack on the SALT 1 Treaty and as an input into SALT 2 then under negotiation, in the form of the 'window of vulnerability' which was claimed to threaten US land-based ICBM forces. Qualitative factors, such as accuracy and vulnerability, rapidly became the language of the moment, and war-fighting scenarios were outlined in which a US President was portrayed as likely to be unwilling to employ his surviving SLBMs because they would entail counter-city strikes, once his more accurate and discriminating Minuteman missiles had been lost to a Soviet counterforce attack. That there was much political special pleading involved in this debate is undeniable and Michael Howard was correct to object to the political unreality of much of the analysis:[21] none the less, the ensuing dialogue was important, less for its substantive proposals than for marking a change of intellectual climate in which the minutiae of nuclear war received unprecedented attention. For good or ill, the waging of nuclear war had become a subject of widespread investigation, even if not universally deemed appropriate for polite academic society. In the public domain, media portrayals of the consequences of nuclear war served to intensify this concern.

If it was barely understandable that the topic of strategic defence should have had little place in elaborations of deterrence-only strategies, it was inevitable that interest in defence would be resurrected in the new climate because no discussion of the normalization of nuclear war could proceed without due consideration being given to that half of the traditional military enterprise.

This is not the place to review President Reagan's SDI in all its complexity. However, there are aspects of SDI which touch directly upon crucial parts of the just and limited war traditions. If for no other reason, there is at least an ostensible paradox in maintaining that the only just cause of war is that of national defence, at the same time as national strategies make no effort to operate in the realm of defence against the enemy's most deadly nuclear weapons. Just as the cause of self-defence has an intuitive just war appeal, so the defensive mode of

---

[21] 'On Fighting a Nuclear War', *International Security*, Spring 1981.

military action can appear to be preferable, in moral terms, to offensive forms of deterrence. On the other hand, such a moral hierarchy has been dismissed as 'pernicious',[22] deterrence being deemed the most effective form of defence in the nuclear age.

Judgements about SDI, in just or limited war terms, must vary in accordance with the fundamental interpretations of that programme that are currently in circulation. It has been presented as a radical attempt to eliminate nuclear weapons and the practice of deterrence based thereon; as the missing dimension of a credible war-fighting capability and hence as a requirement for a more effective nuclear deterrent; as a programme that will lead to the protection of civilians and finally remove the immoral condition of vulnerability in which civilians have been held in the nuclear age; as a defence for missile silos in order to remove the attractions of pre-emptive strikes and hence stabilize the strategic balance; and as an aggressive adjunct of first-strike intentions offering a capability to withstand what remains of the enemy's retaliatory blow.[23]

Assessment of SDI, in other words, depends very much upon the intentions behind the programme, and upon its incidental consequences whether intended or not. On the face of it a strategic development which would both protect civilians and reduce the total destruction resulting from nuclear war would seem desirable from either a just or limited war perspective. If, in doing so, it could have beneficial strategic consequences for crisis and arms race stability, as well as for arms control, it would be doubly welcome. Limited strategic defences have also been positively regarded as a potential check upon escalation, offering a useful fire-break against limited attacks and hence reducing the likelihood of any use of nuclear weapons.[24]

As against this, ethical objections to strategic defence range across the spectrum. At the one end, it is castigated for vitiating the role of nuclear weapons either in just prosecution or in just deterrence of war and, in consequence, for making more likely the onset of conventional wars in which millions of lives may be lost. At the other, it is denounced not for its defensive credentials but for its hidden offensive purpose and destabilizing effects:

[22] McNamara and Bethe, 'Reducing the Risk of Nuclear War', p. 46.
[23] From the vast literature, S. E. Miller and S. van Evera (eds.), *The Star Wars Controversy* (Princeton University Press, 1986) and the Office of Technology Assessment reports in *Strategic Defenses* (Princeton University Press, 1986) are helpful overviews of these competing interpretations.
[24] J. Lodal, 'Deterrence and Nuclear Strategy', *Daedalus*, Fall 1980, 167.

instability can arise if both sides acquire moderately effective nationwide defences. This gives rise to the so-called 'ragged retaliation' problem. That is, with moderate levels of defence and partially vulnerable offensive forces, both sides may have an incentive to strike first in the expectation that the adversary's retaliation, already impaired by the counter-force attack, will be unable to penetrate the defences.[25]

In summary, the tortuous logic of deterrence theory, and its seeming capacity to lend credence to contradictory conclusions, makes it premature, if not impossible in principle, to reach a definitive judgement about either the intentions, or the consequences, of a programme of strategic defence. This, in itself, may be symptomatic of the conceptual shift which nuclear war entails; it is difficult to envisage a defensive programme in the past which would have elicited such ambivalence as to its conformity with just and limited war precepts.

Central to any assessment of the waging of nuclear war is the process of escalation. There are few, if any, commentators who would argue in support of the legitimacy of unrestricted use of nuclear weapons: at best, the use of some types of nuclear weapons, in certain ways, is accepted. How, then, are those categories of legitimate use related to other forms of usage which may result from dynamic processes of escalation?

Escalation is not foreordained. It is not technologically impossible to devise limited forms of nuclear action, nor can we be certain what the politico-military response of command authorities to such action might be. Some, indeed, maintain that the most likely effect of the use of any nuclear weapons is a cessation of hostilities: 'By far the most likely outcome', Luttwak attests, 'is that a war would end very soon if any nuclear weapons, however small, were actually to be detonated by any side on any target.'[26] These are matters for judgement that must remain uninformed by precedent or historical experience.

If just and limited war must concern itself with sensitive judgements about the consequences of military action, and with establishing due proportion between means and ends, it is not self-evident that predicting the escalatory consequences of nuclear action is, in principle, radically different from such assessments in the past. The sharp difference exists in terms of the penalties resulting from an erroneous assessment. However, the abandonment of all risk would, as

[25] D. Wilkening et al., 'Strategic Defences and First-Strike Stability', Survival, Mar./Apr. 1987, 161.

[26] E. N. Luttwak, 'How to Think About Nuclear War', Commentary, Aug. 1982, 25.

Nye suggests, necessitate a sacrifice of other values in favour of survival, a choice that current societies are not instinctively prepared to favour.[27] Whether, in making such choices, one is entitled to put at risk the survival of third parties as a result of environmental damage is another matter.[28] It is a moot point how strenuously nuclear belligerents would be obliged to respect the right of non-combatant immunity enjoyed by those other states not party to the conflict.

There is artificiality in discussing these matters wholly outside the context of strategies of deterrence as the actualities of nuclear war are regarded to be at least one step removed from human experience, inhabiting the shadowland that would become real only if there were a prior failure of deterrence itself. However, by insisting that deterrence and war preparation must be more closely integrated with each other, recent nuclear strategies have themselves contributed to the conflation between shadow and substance that has taken place.

Strategy and Weapons

The attempt to distinguish between various individual weapon systems, as to their conformity with, or violation of, the rules of war, might conceivably be based on a number of considerations. For instance, it might seem appealing that, as just war doctrine has made a separation between wars which are justly defensive and those which are unjustly aggressive, this might be carried over into the weapons themselves with the result that we should seek to prohibit the use of offensive weapons while condoning the use of defensive ones. But this has at least the two problems of rendering difficult those just wars which would require carrying the battle (offensively) to the enemy, as well as the purely practical consideration that almost all kinds of weapons lend themselves to a duality of offensive and defensive functions.

Secondly, the nature of the weapons might be assessed in relation to a principle of discrimination. Since some weapons appear to be less discriminating in their effects than others, and some tend towards all-encompassing effects, they may be objected to on that score. As we already encountered the argument that there may be something immoral about denying the enemy his right to make a moral choice, so it might be contended that those weapons which, by their very nature,

[27] *Nuclear Ethics*, p. 65.
[28] J. S. Nye, 'Nuclear Winter and Policy Choices', *Survival*, Mar./Apr. 1986, 120.

deny us the choice of discriminating between targets, are likewise unacceptable.

Thirdly, it has been common, and is at least imperfectly embodied in the laws of war, that weapons can be categorized into those that are more, and less, humane. Here we are discussing not the manner of their strategic use, nor the legitimacy of the target, but the manner of the suffering which they inflict upon the recipient. According to this criterion, it is in the nature of the wounding, or the nature of the killing, which they inflict that the only sensible distinction between weapons can be drawn.

Any effort to regulate the use of weapons in war therefore comes up against this initial difficulty of articulating the principle upon which it shall proceed. But the difficulties go beyond this. Indeed, some have thought that the effort to humanize the weapons of war is fundamentally misconceived because it can only have the unfortunate consequence of making war that much more tolerable to all and that much more likely as a result. Humane warfare, on this reasoning, is subversive of the goal of reducing, and finally eliminating, war itself. This is to say no more than that the effectiveness of deterrence may actually be proportional to the inhumanity of the weapons likely to be used. Again, which is the worse, a life taken cruelly in an unavoidable war or a life taken humanely in a war that might have been otherwise deterred? Perhaps the antithesis is a false one but it serves to illuminate some of the moral ambiguities which surround the issue.

Another possible snare lies in wait. It concerns the tangible effort to restrict weapons through the laws of war. While this intent is an honourable one, it has been chastised by some, notably Richard Wasserstrom,[29] for deflecting the discussion away from the proper area of concern. According to this argument the restriction of weapons through legal means leads to an overly narrow discussion about which weapons are legal or not, rather than about which weapons are moral or not. This, to Wasserstrom, is the moral cost of the enterprise and it is a price he is reluctant to pay. Because positive international law can only embody that to which the states give their consent, and since there are some weapon systems which states are unwilling to abandon (e.g. nuclear weapons) the debate becomes caught up in the narrow issue of legality whereas it should properly be addressed in the wider terms of

[29] See e.g. his *War and Morality* (Wadsworth, 1970) and 'The Relevance of Nuremberg', in M. Cohen *et al.* (eds.), *War and Moral Responsibility* (Princeton University Press, 1974).

morality. The cost, in other words, is that the law of war pertaining to weapons is itself morally incoherent as it is arbitrary as to which weapons it prohibits and which not, even if the effects of these weapons are broadly comparable, or the effects of some legal weapons the more inhumane.

The Need to Regulate Weapons

It is by no means universally accepted that use of certain classes of weapons should be restricted or prohibited. For instance, Walzer, amongst others, has argued that such restrictions may be desirable in terms of just conduct of war but they are not self-evidently essential to it:

Rules specifying how and when soldiers can be killed are by no means unimportant, and yet the morality of war would not be radically transformed were they to be abolished altogether. . . . Any rule that limits the intensity and duration of combat or the suffering of soldiers is to be welcomed, but none of these restraints seem crucial to the idea of war as a moral condition.[30]

On what bases, therefore, might we seek to distinguish between just and unjust weapons?

1. In accordance with the principle of discrimination, whichever specific form it might take, it might be considered necessary to have weapons that are capable, in their effects, of respecting it.

2. In accordance with the principle of proportionality, it might be considered that any weapon which, by its intrinsic use, is likely to exceed in destruction the value of the object which is being sought, cannot be an acceptable weapon of war since it would be a contradiction to have means which are incompatible with the desired ends.

3. In accordance with a principle of humanity, it might be urged that weapons should not be employed which inflict unnecessary suffering upon the persons against whom they are used. Indeed, this notion may be regarded as partly derivative from the principle of non-combatant immunity. According to this view the use of a weapon is necessary to incapacitate militarily the enemy combatant: it is directed at the soldier *qua* combatant and not at the soldier *qua* human being. As soon as the combatant is incapacitated, he is by definition no longer actively bearing arms and is entitled to the immunity accorded by his

---

[30] *Just and Unjust Wars* (Basic Books, 1977), 42.

new status. Any additional suffering or pain which the weapon inflicts upon him is gratuitous and strictly superfluous to military necessity. On this criterion some disabling chemical agents, without long-term effects, might be deemed more humane than the bullet.

4. Combining categories 1 and 3, it might be worthwhile to distinguish between those weapons which fail to discriminate but are relatively humane in their effects, and those which fail to discriminate and do so inhumanely. A hail of conventional fire which catches civilians in its midst, or bombardment of a city in siege by conventional artillery, might serve to illustrate the former: the wiping out of an entire population by the spread of a deadly disease would illustrate the latter. Are both equally unjust in that they infringe discrimination, or is the latter doubly objectionable in compounding its indiscriminateness by its inhumanity?

5. Latterly, there has been an increased tendency to seek to prohibit specific weapons on account of their indiscriminateness in a much broader sense than that traditionally employed. According to this, weapons should be proscribed which threaten the human environment, and therefore the interests of all humanity, including the interests of succeeding generations within this term. Hitherto there has scarcely been need of such a prohibition as there was no individual weapon which had the capacity to render such effects. It is interesting, therefore, that this principle seems to have been one previously unstated, but now articulated as a response to nuclear weapons.

6. Combining categories 2 and 5, and bearing in mind the just war injunction that the object of war is to restore a more just peace, it could be claimed that a test of proportionality, or of the long-term effects upon humanity, might be whether or not the weapon allows for the restoration of such a peace.

Issues of Principle and Practice

The discussion about the morality or legality of individual weapon systems raises a number of important practical and philosophical issues. Not least amongst these is the central point of debate as to whether we can, in any case, associate moral qualities with inanimate objects such as weapons. In the view of some, the attribution of moral qualities to the weapons is a philosophical nonsense, as in the opinion of Cohen:

Military weapons by themselves can hardly be classified in moral terms. . . . [I]f one is to make moral determinations regarding military weapons, it has to be in the context of how these weapons are used and to what purpose.[31]

The weapons themselves, on this reckoning, are morally neutral: the moral discussion can properly be conducted only in connection with the human agents who employ them and is about the purposes of such use. What this seems to leave out of account is the extent to which it may be said that some weapon systems pre-empt moral choice on the part of their users. Surely if it could be agreed that a particular weapon could never be used in a morally acceptable way, any discussion about the agent's purpose becomes redundant. Whether or not it is philosophically accurate to speak of such a weapon as being immoral, the practical import is the same at the end of the day. However, leaving this objection aside, the thrust of the argument for the value-neutrality of weapons is to make the discussion about human uses of them the only valid distinction, overriding completely the physical and technical properties of the weapons themselves.

Thus, it has been concluded by some that to distinguish between weapons, as weapons, is to make an arbitrary distinction, lacking in philosophical substance, and leads to the creation of artificial thresholds. In the words of Johnson, 'armaments should be judged morally not in terms of whether or not they are nuclear but rather in terms of their intended use in war'.[32] By this reasoning, the threshold between nuclear and non-nuclear weapons is a morally neutral one.

What this argument fails to consider is the suggestion, previously encountered, that there may none the less be some moral value in thresholds which are themselves morally arbitrary if, in their absence, no other thresholds are available. This takes us back to the issue of saliency and to the notion, adapting Lenin's aphorism that 'quantity has a quality all of its own', that 'saliency has a morality all of its own'. The choice which confronts us is a stark one, between Wasserstrom's critique of the moral costs of the laws of war and the alternative moral costs of not having restraints which, even if morally incoherent, are restraints none the less in an activity in which any practicable restraint is a scarce and valuable commodity.

If the subject of the morality of weapons is to be considered only in terms of their putative use, two further matters command our

[31] S. T. Cohen, 'Whither the Neutron Bomb? A Moral Defense of Nuclear Radiation Weapons', *Parameters*, June 1981, 19.
[32] *Can Modern Wars be Just?*

attention. When we say that the issue of morality hinges on the nature of use, this passes over in silence the matter of possession. This presumably comes first in the order of logical priorities and, even were it conceded that there can be no intrinsic immorality in a weapon, possession is presumably a human act. If weapons themselves are value-neutral, and if possession is not self-evidently a goal-oriented activity, can we make any kind of moral judgement about the possession of weapons?

But possession might be a kind of use? At this point the debate about the use of weapons and about their possession collapses into the wider issue of the relationship between use and threatened use, or between use and deterrence. This is a subject to be discussed further below.

A final point of philosophical principle merits attention. Let us accept for the moment that nuclear weapons are unjust weapons for whatever combination of reasons listed earlier. What, then, is the relationship between a class of clearly recognized immoral weapons and other weapons, perfectly moral in themselves, but the use of which would, with reasonable certainty given the framework of strategic doctrines, lead to the employment of the unjust weapons as well? What prompts the question is clearly the concept of 'escalation' and its prominence in almost all discussions of nuclear deterrence and, indeed, its strategic institutionalization in such postures as that of flexible response.

Can we class a weapon as immoral if it has the potential for escalating hostilities which will culminate in the use of unjust ones? By analogous reasoning, if we were to condemn the scale of the violence perpetrated during the First World War, the pistol which killed the Austrian Archduke would be a very immoral weapon indeed.

The case is more compelling in the present strategic circumstances. Were the use of conventional weapons to trigger the use of nuclear weapons, this latter would not represent an accidental and unforeseen eventuality. Rather is it part of the integrated strategic design which seeks to make the use of conventional weapons unlikely by tying the use of these conventional weapons to the almost certain consequential use of nuclear weapons. If escalation is thus institutionalized, how are we to make tenable distinctions between weapons if the use, and non-use, of all weapons is so intimately, and so inextricably, interwoven: if the use of one is immoral, so surely must be the use of any?

As already argued, neither just war nor limited war tradition can accept such a conclusion. Its thrust is to deny the prospect of limited wars and to make impossible the fighting of just ones. Can one weapon have such power as to destroy both traditions simultaneously? The reality of the problem is, naturally, denied by those who do not accept the initial assumption that nuclear weapons *are* immoral. A number of writers have accordingly argued that nuclear weapons are susceptible to the same moral distinctions as most other kinds of weapons and have sought to delineate which kinds of nuclear weapon might best conform to the traditional laws of war. Some have found virtue in the neutron bomb. Others would distinguish generally between tactical nuclear weapons and other kinds. James Johnson has recommended the cruise missile as a nuclear weapon that can be reconciled with just war theory:

If noncombatant immunity . . . is to be taken seriously in our thinking about war, the cruise missile represents a definite positive development in weapons systems. After nearly a generation of strategic thought that never strayed far from the idea of direct attacks on enemy population centers, the cruise missile opens up the possibility of devising a strategy based on attacks that are aimed not at destroying large numbers of civilians but at neutralizing the enemy's military power. . .[33]

The argument is, to a degree, simplistic. If Johnson does not allow that morality is an integral characteristic of an individual weapon, is he not equally in error in believing that a specific strategic mission, let alone a defined target set, is inherent in any one weapon?

Nor is it at all clear that this approach to the issue does not come up against a similar problem of escalation. Even if the issue is not that of nuclear weapons as a rigid category, being related by the prospect of escalation to other categories of weapons, does there not still exist the danger that certain moral uses of nuclear weapons, if such they be, will escalate to include immoral uses as well? On the principle of saliency, it could be said that escalation is that much more likely to occur as the belligerents will certainly be more impressed by the breach of the technical threshold (the use of nuclear weapons) than by the retention of the moral one (just use of nuclear weapons only).

One point remains to be made on the morality of weapon systems. How much is 'justified' by appeal to military necessity? If we are yet persuaded that there may well be circumstances in which, for reasons of strict military necessity, even dubious weapons must be resorted to, it matters considerably what precisely is the nature of the exoneration

[33] Ibid. 124–5.

which is made. On the one hand, we can adopt the form of extreme dispensation and argue that military necessity makes 'right' what would otherwise have been 'wrong'. On the other, we argue the weaker line that military necessity explains what we have done but does not justify it: what is wrong to do remains wrong even if required by military necessity but can be mitigated by some appeal to the 'lesser of two evils' argument: for instance, the wrong done by using the weapon is less than the wrong which would have resulted, perhaps by losing the battle to a barbarous opponent, by refraining from its use.

How readily such general contentions about weapons might be applied to nuclear weapons remains an open question. Much of the debate hinges on acceptance of the 'uniqueness' of these weapons or otherwise. It must certainly be the case that, if valid, the extreme versions of the nuclear winter scenario would establish the singularity of nuclear weapons beyond any shadow of doubt. At the moment, however, the status of this theory remains under challenge.

What further complicates the discussion of nuclear weapons is that their virtues are seldom considered in terms of actual military usage but rather as instruments of deterrence. In what respects, if at all, this change of focus affects our assessment will be considered in the final chapter.

6

CONTEMPORARY NUCLEAR STRATEGY
AND WAGING PEACE

WHAT we have thus far been considering has been a variety of philosophic issues thought to touch upon the matter of the waging of war. We are now required to extend the discussion by enquiring how far these selfsame principles are relevant to the waging of peace. Is there a *ius ad pacem* and a *ius in pace*, the counterparts of the just war theory, that govern those military measures that we take, as part of deterrence, not only to avoid a hot war but also to manipulate the cold peace? If it is true, as deterrence theory claims, that preparation for war is a necessary precondition of the maintenance of peace, how should the restrictions on war be applied to those peacetime preparations?

The notion of deterrence is as old as the history of human warfare but it was only with the nuclear age that detailed consideration came to be given to the nature and conditions of deterrence in separation from the discussion of the fighting of war. Hence a bifurcation occurred and the analysis of military strategy followed a path that did not always head in the same direction as that of the analysis of war avoidance. Deterrence, a necessary adjunct of armed force, was liberated and developed into a self-contained theory of its own.

Not only was nuclear strategy different because of the emphasis that it placed on deterrence: what was unique about the nuclear age was that this deterrence was divorced from defensive capabilities strictly understood and accordingly the threat held out to the enemy was no longer that of the punishment that would be endured in trying to breach a defensive perimeter but rather the societal punishment that would follow upon any aggressive action. For such deterrence to be successful, however, the threatened retribution must be credible and, unlike conventional forms of deterrence where the threat resides in the military capability of armed forces, the credibility of nuclear punishment has been increasingly regarded as an artefact, something to be created and which needs constant fine tuning to respond to changing circumstances.

Deterrence theory has not confined itself to analysis and description. Additionally, it has had a self-assigned task of prescription, addressing value judgements about the goals of society and about the appropriate risks to be taken in their pursuit. Arguably, the psychological framework of deterrence has also influenced behaviour.[1]

Where then are we to locate deterrence on the spectrum covering the waging of war and the waging of peace? It was earlier contended that just and limited war doctrines occupy a middle ground between the extremes of restraint and permissiveness, that is, between pacifism at the one end and 'war is hell' at the other. Ostensibly, deterrence also is a middle position between renunciation of war and the horrors of nuclear conflict. It is precisely because it is so that Mandelbaum predicts its continuance into the medium-term future: 'the alternatives, disarmament and war, are either too difficult to achieve or too terrible to risk'.[2] If both deterrence on the one hand, and just and limited war on the other, have their inspiration in the common ground of avoiding the poles of non-resistance and of unbridled violence, can we validly equate the two? Might it be argued that deterrence is the current embodiment of the two traditions and if just and limited war have traditionally required that wars be fightable—albeit exceptionally— whereas deterrence requires war's avoidance, this is but a reflection of the deep-seated change that nuclear weapons have wrought?

There are intimations in the literature, even if implicit, that the transition from just and limited war to deterrence is by no means as revolutionary as some have thought. Colin Gray, who more than most has argued for the accommodation of nuclear weapons to conventional strategic concepts, suggests that just war theory and deterrence share a common preoccupation with the idea of proportionality:

Proportionality, as well as constituting an important element in the doctrines of *jus ad bellum* and *jus in bello* of the Catholic Church, plays a significant role in the strategic reasoning of deterrence theorists and practitioners. Threats that manifestly are disproportionately large to the scale of hypothetical offense are deemed to be incredible and hence ineffective.[3]

The use of common words, however, does not amount to the speaking of a common language. It is evident, in this case, that Gray's analysis

---

[1] See R. Jervis, *The Illogic of American Nuclear Strategy* (Cornell University Press, 1984), 37–40.

[2] M. Mandelbaum, *The Nuclear Future* (Cornell University Press, 1983), 121.

[3] C. S. Gray, 'War Fighting for Deterrence', *Journal of Strategic Studies*, Mar. 1984, 15.

shares no philosophical ground with just war theory. The former says no more than that a disproportionate deterrent threat will not work: the latter proclaims that a disproportionate act must not be done. Nor, if we are to believe its critics, can deterrence be reconciled with the prescriptions of political limitation in war. The intention of deterrence may well be to avoid extreme forms of war but, to be successful, it must threaten the execution of precisely such a war, and it cannot infallibly guarantee that such war will never take place. It is at this point—not on the basis of stated aims but because of adopted means and potentially unintended consequences—that, according to Booth, deterrence parts company with the tradition of politically instrumental war:

Since we cannot guarantee that the nuclear leviathan will always deter, even if it usually will, then it is obvious that the un-Clausewitzian notion of nuclear war is integral to the idea of nuclear deterrence. It must therefore be concluded that nuclear deterrence is ultimately un-Clausewitzian. A strategy which has no reasonable answer to its breakdown, except inviting the prospect of 'absolute war', cannot be considered to be Clausewitzian in its inspiration, however remote the possibility of that breakdown.[4]

Few aspects of the philosophy of contemporary warfare have received as extensive consideration as has deterrence. In its essentials, the debate about the moral status of deterrence, and about its conformity to requirements of just and limited war, seems to centre upon the very complex interrelationship between the threat on which deterrence is based and the execution of that threat if the deterrent were to fail. Possibly there has been too widespread a disjunction between the two realms of threat and execution: only in terms of such a radical separation can we fully comprehend how those traditionally wedded to a posture of deterrence-only by means of a threat of assured destruction have thereby implicitly endorsed a strategy which, in execution, would entail a palpable violation of principles of both discrimination and proportionality; at the same time, those who have espoused nuclear war-fighting strategies which would go some way, at least in principle, to respecting the just war code are those whose moral integrity has been most vehemently impugned. This is but one of the many striking paradoxes in which the debate about nuclear

4 K. Booth, 'Unilateralism: A Clausewitzian Reform', in N. Blake and K. Pole (eds.), *Dangers of Deterrence: Philosophers on Nuclear Strategy* (Routledge & Kegan Paul, 1983), 52.

deterrence abounds and, on this evidence alone, the situation must be regarded as highly complex.

Evolution of Deterrence

Although nuclear deterrence has been a major element in Western strategies since 1945, it is misleading to assert continuity and consistency in such a posture. Rather, various policies have underlain the strategy of deterrence and there has been a number of significant watersheds in its evolution. The following does not pretend to be a comprehensive history but merely an indication of what has been practised as deterrence: it is necessary to have some clear perception of this before the concept of deterrence can itself be meaningfully analysed.

Initially the US posture of deterrence was latent rather than explicit. No specific nuclear doctrine was articulated in the late 1940s, reflecting both the tardiness of the atomic weapon production programme and also lack of clarity about the military functions of atomic weapons. Such nuclear deterrence as existed was, in any case, unilateral reflecting the monopoly that the United States enjoyed for the first few years of the nuclear age. One of the earliest paradoxes was to be that it was only when the monopoly was broken that the United States began fully to integrate nuclear weapons into national strategy and to declare publicly a threat of nuclear attack. With hindsight, we can see that some of the problems of credibility of nuclear deterrence were to result from this conjunction.

The early Eisenhower phase of US strategy was dominated by the ambivalent experience of the Korean War and by the fruits of technological innovation. While the outbreak of the Korean War initially called into question the utility of nuclear threats in defending peripheral areas, and stimulated the expansion of conventional forces, the longer-term reaction was to shy away from the financial burdens conventional rearmament entailed and to underline the cost-effectiveness of a nuclear strategy. Technologically, this conclusion was supported by the development of the thermonuclear weapon, which promised an even more frighteningly persuasive deterrent, and the emergence of a new generation of tactical nuclear weapons, in which the West enjoyed an advantage and which promised a nuclear deterrent combined with some capability of denial of Western Europe to Soviet forces. Only subsequently was the full cost of using these weapons in central Europe to be assessed.

Confidence in the massive retaliatory deterrent was already on the wane when Sputnik, and a potential Soviet capability for attacking the US homeland, called it radically into question. The emergence of a NATO posture of flexible response over the next ten years testified to the compromise which the United States and its European allies finally reached, the Europeans insisting on a measure of pure deterrence which would avoid conventional or tactical nuclear war in Europe, at the same time as the United States, by providing a rationale for substrategic forces, bought itself time before the execution of its nuclear guarantee, with the full implications for the American homeland which such action would entail.

It was during this same period that the foundations of nuclear mutuality were laid. From the mid-1960s to the early 1970s, the USSR attained rough strategic parity with the United States and it is this fact which has driven the elaboration of deterrence theory ever since. The various efforts to devise credible deterrent postures by means of limited nuclear war options and countervailing and prevailing strategies have been essentially motivated by the need to escape the dilemmas created by parity—how can one credibly threaten to employ nuclear weapons against an enemy who is comparably able to retaliate against oneself? The very posing of the question leads directly into a discussion of the operational dimensions of nuclear hostilities. Herein lies the problem. 'Nuclear weapons fulfil their function by avoiding war, not fighting it', one writer suggests. 'The tragedy for deterrence is that discussion of its defensive aspects exposes some of its rawest nerves.'[5]

Thus far it has been assumed that one can discuss the evolution of nuclear strategy as if the reality of that strategy were unambiguous. In fact, this is not the case. When we speak of the nuclear policy of the United States we actually refer to a number of disparate and occasionally competing policies. For instance, the public is most aware of declaratory policy, that which is announced as being the strategic intentions of the state. Such declaratory statements may, however, bear little similarity to operational policies, the tasks that would be assigned to US forces in the event of war. As Gray points out, whatever

⁵ H. Strachan, 'Deterrence Theory: The Problems of Continuity', *Journal of Strategic Studies*, Dec. 1984, 400.

operational plans might be, deterrence requires maximizing the anxieties of Soviet planners.[6] The importance of this is twofold. Inasmuch as the core issue, in assessing the relevance of codes of war for the posture of deterrence, lies in the relationship between threat and use, this adds a further layer of complexity in distinguishing the reality of the threat. How are we to judge the reality of the intention to use nuclear weapons when this intention is itself compartmentalized amongst bureaucratic agencies and located at varying distances from the operational sphere? Secondly, the assessment of the ethical acceptability of deterrent threats is frequently believed to be related to the precise nature of the threat in terms of stated targets. Again, the problem confronting the analyst is in knowing what these policies might be as the declared posture of assured destruction, with its implications of wholesale counter-population targeting, seems to be an imperfect description of US operational strategies over the past three decades. In the business of deterrence, what one would actually do in the event need not be what one is threatening to do before the event, a situation already impenetrable enough without the opaqueness of the actuality of the threat, hidden behind declared and operational intentions. In these circumstances, the development of a sensible *ius in pace* for deterrence runs the risk of becoming an exercise in mysticism rather than in practical ethics.

Principles and Practice

The central issues involved in the analysis of deterrence tend to be the relationship between means and ends, between threat and its execution, and, fundamental to all this, an exploration of the concept of intention, conditional or otherwise.[7]

What are the stated ends of a posture of deterrence? According to one representative account, they are twofold: 'Through its possession of nuclear weapons, the Western alliance is able to avoid the two evils of a conventional war and of almost certain defeat through such a war at the hands of the Soviet Union, with a consequential loss of freedom.'[8] To this must surely be added a third. 'Strategic nuclear

---

[6] 'War Fighting for Deterrence', p. 10.

[7] A brief summary of these issues, from a perspective critical of deterrence, can be found in B. Paskins, 'Deep Cuts are Morally Imperative', in G. Goodwin (ed.), *Ethics and Nuclear Deterrence* (Croom Helm, 1982), 97–9.

[8] G. Rumble, *The Politics of Nuclear Defence* (Polity Press, 1985), 210.

weapons', Hockaday insisted, 'are a vital deterrent . . . to the use of similar weapons by an opponent.'[9] In short, the ends of deterrence are the protection of values, the avoidance of costly conventional warfare, and prevention of the enemy's use of nuclear weapons.

To these 'peacetime' goals of deterrence, a fourth 'wartime' objective might be added. As against the notion that deterrence operates only as long as peace is maintained, and is abandoned as soon as the shooting starts, there is a substantial body of theory on the subject of intra-war deterrence. Its essential task is to maintain a proscribed area within war, by holding out threats of punishment if that area is violated.

Do these ends create problems in practice? The summary objections are that the means of deterrence may be corrosive of the ends as the adoption of an intrinsically immoral strategy, no matter the rightness of the end, erodes the value-basis of our own society. Secondly, rather than using nuclear threats to avoid conventional war, we ought properly to move towards conventional deterrents to escape the necessity of reliance upon nuclear strategies. Finally, it has been claimed, deterrence is not useful in the goal of preventing the spread of war, either in an escalatory, geographical, or intra-war sense, precisely because the downward links in the chain intended to avoid the lowest levels of violence also lead upwards to the highest. Aspects of these various objections will be considered below.

In order to outline the issues which are at stake, rather than in an effort to resolve them, it can be suggested that the relationship between nuclear deterrence and nuclear use can be conceived of in four very distinct ways. Each provides an interesting, and different, perspective on the moral status of deterrence and upon its legitimacy as a means of conducting war and peace.

1. The first takes the view that deterrence is to be justified in its own terms because it is in principle different from the actual use of nuclear weapons. Indeed, since the explicit intention is that of not using nuclear weapons, any discussion of use is radically irrelevant to the issue. Whatever the situation regarding the use of nuclear weapons, we must approach deterrence from its own discrete standpoint.

To the extent that this argument is allowed, nuclear deterrence is

---

[9] A. Hockaday, 'In Defence of Deterrence' in Goodwin (ed.), *Ethics and Nuclear Deterrence*, p. 72.

distanced from discussion of the effects of use of nuclear weapons. Accordingly, the criticisms of nuclear weapons *per se*, namely that they infringe proportionality and discrimination, are beside the point since this says nothing about the condition of deterrence itself in which these weapons might be manipulated but are not actually employed.

There are a number of objections to this procedure. The first, derived from some ethical systems, is to deny the moral validity of this separation. In its light, if a thing is immoral to do, it is equally immoral to threaten to do. This stance equates the moral condition of the act with the moral condition of the threat to perform the act. It is itself a questionable position to hold as there is some difficulty in accepting an ontological equivalence between act performed and act conditionally threatened but has the virtue of casting doubt on the 'clean hands' assumption of deterrence theory—namely, that the process of threatening is in no way related to deplorable human outcomes. Since deterrence itself claims credit for beneficial consequences in the maintenance of peace and stability, it seems only fitting that the potential consequences of threatened use should at least be borne in mind. Selective consequentialism is an unsatisfactory ethical position.

Secondly, deterrence does not, in any case, express an absolute intention not to use nuclear weapons; rather it expresses a conditional intention to use them in certain circumstances. This being so, it is an illegitimate procedure to disallow discussion of the characteristics of usage of these weapons. To ignore the threatened, if conditional, use of nuclear weapons is to present a sanitized, but completely artificial and misleading, conception of the reality of deterrence.

Thirdly, if deterrence were as absolutely separated from the prospect of use as this argument implies, it is difficult to see how it could be effective. Deterrence may require only the remotest possibility of use in order to be credible but this remotest possibility is sufficient to disallow the radical separation between deterrence and use which is the basis of this contention. Were there not this inextricable link between credibility and the prospect of use, it would be extremely difficult to understand why nuclear strategy in the past fifteen years has moved so comprehensively in the direction of adoption of limited and controlled, and thereby more credibly performable, operations. Since deterrence theory itself integrates the act and the threat into its own logical structures, this provides compelling evidence for the moral theorist to do likewise.

2. Accepting that deterrence and use are related, we might initially

posit that the justification for deterrence resides in the fact that use of nuclear weapons could be justified in some circumstances. According to Okin, this is in fact one of the assumptions underlying the argument put forward by the Catholic Bishops in the United States:

the bishops' acceptance of the moral equivalence of intention and action means that their only way of arriving at the conclusion that nuclear deterrence is not evil is to argue that, in certain circumstances, certain *uses* of nuclear weapons are morally acceptable.[10]

This is a significantly different manner of argument. Rather than emphasize the moral disjunction between threat and deed, moral equivalence between the two is assumed as part of the procedure of discussion. Indeed, the legitimacy of deterrence is perceived to be secondary to the fact of the legitimacy of nuclear employment itself. Since there is no intrinsic objection to the use of nuclear weapons (provided always that other just war criteria are satisfied), there can likewise be no objection to their threatened employment.

This simply pushes us back to the earlier discussion about the possibility of a just nuclear strategy in execution. The *ius in pace* is contingent upon the existence of a nuclear *ius in bello* and can intend to do only that which it is permissible to do as a legitimate act of war.

This is a perfectly plausible, but by no means so common, line of analysis. It is more frequently contended in the public debate that nuclear deterrence is undesirable but gains its conditioned acceptability from its beneficial consequences and from the lack of any viable alternative. None the less, it is by no means self-evident that the actual use of nuclear weapons is illegitimate. On the contrary, the structure of the vindications of the use of the atomic bombs in 1945 is such as to suggest the opposite, and the lingering impression that this first use of atomic weapons was itself morally acceptable has had its impact on the ensuing debate. As I have argued elsewhere,[11] it is misleading to believe that we have tolerated nuclear weapons because of the benefits derived from deterrence: on the contrary, we have learned to live with nuclear deterrence because, at heart, the events of 1945 immediately established the moral legitimacy of using nuclear weapons. To that extent, the Catholic Bishops are simply tapping a well-established tradition.

---

[10] S. M. Okin, 'Taking the Bishops Seriously', *World Politics*, July 1984, 531.

[11] I. Clark, *Nuclear Past, Nuclear Present: Hiroshima, Nagasaki, and Contemporary Strategy* (Westview Press, 1985).

3. The inversion of this relationship leads to the opposite conclusion that the legitimacy of usage can be derived from the moral requirements of deterrence. For our immediate purpose this is itself a digression because it seeks to shed light on the moral issue of nuclear use while offering no evidence for the assumption of the morality of deterrence. However, it is interesting to note that such a relationship has been posited. For instance, Richard Brandt has contended that 'a deterrent is of great importance; having the deterrent requires willingness to employ it; therefore, employment is morally justified in the total circumstances'.[12] This is less revealing for what it says about the source of the moral purposes of deterrence than it is about the integral relationship between deterrence and use. If just deterrence can be derived from just use, can we not similarly derive just use from just deterrence? The latter is certainly the less usual formulation of the relationship. We have traditionally found it easier to accept that it is right to threaten what it is right to do than it is to accept that it is right to do what it is right to threaten. What this argument tells us, if accepted, is that the *ius in bello* is derived from the *ius in pace* and not vice versa.

Of the various interrelationships between threat and use, this is the most logically puzzling. One can readily comprehend, if not necessarily accept, the argument that it is wrong to threaten what it would be wrong to do. Likewise, it is understandable that some should argue that it might even be acceptable to threaten what would not be permissible to do if the beneficial consequences of the threat were such as to warrant it.

Brandt's formulation is, however, a radical one. It seems to deduce the ethical status of the act from the ethical necessity of the threat. As such, there is a serious chronological confusion involved. The credibility of the threat is certainly contingent upon the enemy's belief that the punishment will be performed. If, however, the moment for infliction of this punishment arrives, the deterrent has already presumably failed and the question of what it is morally preferable to do must surely then remain an open question and cannot be determined by the requirements of the precedent, and now expired, threat. The only lingering value of executing the threat might be in the context of future situations if it is deemed necessary to establish the seriousness of one's deterrent purpose but this extends the argument

---

[12] R. Brandt, 'When is it Morally Permissible to Use Tactical Nuclear Weapons?', *Parameters*, Sept. 1981, 78.

beyond the immediate terms of Brandt's discussion. Brandt's proposition remains a confusing and unsatisfactory one.

4. In the interests of logical completeness, a final situation must be mentioned. This is the argument which again tells us little about deterrence except to insist that nuclear usage is legitimate in its own terms and does not depend upon any prior discussion of deterrence. Its practical outcome is similar to that postulated in position 1 above but its philosophical starting-point places the emphasis elsewhere—not upon the uniqueness of deterrence but upon the moral self-containment of nuclear use itself. However, its implications for deterrence are not difficult to discover. If it is the case, as the position tends to hold, that nuclear use is morally unproblematic, it would be hard to imagine any circumstances in which a threat of such use could be morally disqualified.

The above perspectives provide much of the intellectual substructure upon which the specific debate about the ethics of deterrence has proceeded. This debate has centred upon the relationship between threat and use, as well as upon the consequences of either or both.

The main form of the consequential proposition has been that deterrence is acceptable 'because it has worked'. Given that nuclear weapons are part of the contemporary reality, they have to be integrated into contemporary strategies in some form and, in the most optimistic assessments, it is the beneficial effect of deterrence that has contributed to the forty years of nuclear peace. Set against this is the counter that the case is unproven: deterrence, to be successful, must work always as the costs of its failure would be exorbitantly high. But neither is this the last word in the argument. If the view that nuclear deterrence has prevented wars that would otherwise have broken out is taken seriously, we are confronted with a complex moral calculus. The dangers of a failed deterrence are only a part of the analysis of moral consequentialism, such that, even if deterrence is wrong, it may be the lesser evil and may well serve the moral purpose of saving as many lives as it conditionally threatens. This is the strident conclusion of Luttwak:

By what doctrine of theology, by what theory of morality, by what rule of ethics is it decreed that the small risk of nuclear war is a greater evil than the virtual certainty of the large-scale death in great-power wars no longer deterred?[13]

---

[13] E. Luttwak, 'How to Think about Nuclear War', *Commentary*, Aug. 1982, 28.

The debate also involves the technical issues of whether or not such nuclear usage is attainable as might conform to recognizable just war principles. Since the case both for a more credible, and a more just, nuclear strategy has been predicated on military capabilities that are significantly untested in operational conditions, the hypothetical nature of these discussions is further intensified but none the less forms an important element of the overall ethical assessment.

Finally, given the duration of the period to be covered by deterrence (potentially for ever), it raises the question of whether such wartime considerations as supreme military emergency can be appealed to on a timeless basis. It was, for instance, Walzer's contention that the strategy of area bombing might have been acceptable as performed by Britain in 1940–1, standing alone in the prosecution of a just war and with no other means of military action available.[14] A strategy that would in other circumstances be illegitimate comes to be tolerable in a situation of supreme national emergency. The problem lies in relating such an analysis to an ongoing relationship of nuclear deterrence. What are we to make of the concept of infinite supreme emergency and how does this shape our ideas about how we might legitimately wage peace into the indefinite future?

Practical Issues of Deterrence

To conclude this review of deterrence, two final issues will be considered. These are the questions emerging from the first-use debate and, secondly, the ethical issues pertaining to the strategy of extended deterrence. If deterrence in general can be discussed within just and limited war frameworks, so presumably can these specific mechanisms of deterrence.

First Use

A just war or limited war discussion of first use of nuclear weapons proceeds through most of the general arguments about the nature and characteristics of nuclear weapons, such as their indiscriminateness and lasting after-effects, which have already been outlined. Similarly, the issue of first use poses afresh the dilemma in assessing threat as against execution in moral terms. This section will confine itself to the

---

[14] *Just and Unjust Wars* (Basic Books, 1977), 255–63.

specific issues posed by a decision to employ nuclear weapons first. The principal objections to such a decision would be of the following kind. As its name implies, first use cannot be presented as retaliation in kind. Thus those vindications of the possession of nuclear weapons merely to prevent their employment by the enemy, or for use as a reprisal against their use by the enemy, do not hold.

Secondly, and relatedly, first use represents the breaching of a threshold. A policy of first use to that extent both undermines any notion of the absolute distinction between nuclear and non-nuclear weapons and also erodes the value of a salient distinction between them even if that distinction is not accepted to be an ethically significant one. More exactly, it might be said that it threatens that the crossing of one threshold (any form of aggression) entails a reasonable likelihood of the crossing of a second (nuclear response).

Thirdly, and flowing directly from this, first use of nuclear weapons entails a threat to any conception of limited war because it has breached this threshold. It does not follow that escalation thereafter becomes irreversible but most analysts would concede that the use of nuclear weapons would apply immense pressures of an escalatory nature. This was an important part of the argument of the 'Gang of Four' who urged of the separation of conventional from nuclear weapons that maintaining this 'firebreak wide and strong is in the deepest interest of all mankind'.[15]

Fourthly, the escalation represented by first use, as policy, cannot be characterized as a mechanical or dynamic escalation, as more a product of synergistic interaction than of rational choice. To declare an intention to employ nuclear weapons as a conscious act of national policy is to threaten to become involved in a process which will lead to the use of these weapons by choice and not by mindless escalation.

Fifthly, it might be considered that the enunciation of a first-use policy is provocative, guaranteeing that the enemy will plan accordingly and base his own strategic conceptions on the early use of nuclear weapons. Indeed, it may well be that the knowledge that theatre nuclear weapons will be used by NATO forces early in a war would encourage the USSR to pre-empt against NATO nuclear stores and against airfields.

Finally, in its emphasis on avoiding all war, it could be objected that a first-use policy undercuts that use of force which is sanctioned by

15 McG. Bundy, G. Kennan, R. McNamara, and G. Smith, 'Nuclear Weapons and the Atlantic Alliance', Foreign Affairs, Spring 1982.

both just and limited war traditions. If, as earlier suggested, both are traditions for the actual waging of war, there is obviously some difficulty in reconciling them with an idea of deterrence, the principal virtue of which is to avoid all forms of war altogether. In this perspective first-use deterrence loses on two counts. It loses if deterrence fails and an unacceptable nuclear war breaks out: but it loses also if all war is prevented, including those wars that are both legitimate on just war grounds and deemed to be politically necessary as well as proportional. In seeking to prevent the former, first-use deterrence stands in the way of the allowable uses of force.

As against all this, the proponents of first-use strategies present a variety of arguments which basically derive from two sets of assumptions. The first is that, as a system of deterrence, our judgement of it must focus on what it is designed to prevent and not upon what might occur if the threat fails.[16] The only real intention is to prevent hostilities. Secondly, in assessing how just or limited first-use threats might be, the point of comparison is not with a condition of peace but with other warring alternatives. Deterrence, in this vision, is not an alternative to stability but to conventional war.

The specific contentions derived from these assumptions can be quickly summarized. According to the critics of no-first-use policies, conventional deterrence has demonstrated its inadequacy by historical precedent and, additionally, the current conventional balance is such as to encourage, rather than prevent, aggression. Moreover, once a conventional war breaks out, there could be no subsequent guarantee that nuclear war would not follow. Given this risk, it is better for all parties to be aware of the dangers at the outset, and to minimize the likelihood of any conflict erupting, than to stumble into nuclear war unthinkingly.

Secondly, given that nuclear weapons exist, it is wise to assume that they will be used and once this assumption is made, the credibility of no-first-use declarations is called into question. As with all limited war ideas, this particular proposal faces a historical challenge in believing that either belligerent would be prepared to lose at a conventional level even when a nuclear arsenal was available to it.

Finally, it can be argued that far from being antithetical to the idea

---

[16] There is a qualification to this. Some proponents of first use argue the case for military reasons, in that the threat of nuclear use induces enemy dispersal and restrains massing of forces necessary for an effective offensive. See e.g. B. Rogers, 'The Atlantic Alliance: Prescriptions for a Difficult Decade', *Foreign Affairs*, Summer 1982.

of just or limited war, the threat of first use of nuclear weapons is designed to enforce restrictions in war, not to lead to unlimited hostilities. For instance, in the mid-1950s, the theory of graduated deterrence was propounded in Britain. According to this theory, the function of the H-bomb was to serve as the ultimate regulator which would compel belligerents to abide by various tacit conventions of warfare as regards targets and size of weapons. The role of the ultimate deterrent in this conception is not the prevention of war but the prevention of a worse kind of war than that which has already erupted.

However, it must be said that this last argument is not as commonly deployed now as it was in the earlier period. Because of fears about escalation from any kind of hostility, the emphasis in the first-use analysis in recent years has been upon the avoidance of war altogether rather than upon trusting to the deterrent to contain hostilities that have already broken out. However, the fact that this argument has been used in the past indicates that there is perhaps no intrinsic incompatibility between the idea of first use of nuclear weapons, as a deterrent to total war, and the just and limited war traditions.

Any discussion of nuclear first use cannot avoid the related subject of conventional deterrence because '[t]he logical consequence of the much debated no-first-use principle has always been a requirement to strengthen conventional forces'.[17] This is a topic that has been widely canvassed in recent years, particularly as a consequence of the perceived advantages of new conventional technologies, both as defensive force-multipliers and as instruments for effectively executing missions traditionally believed to be possible by nuclear means alone. Much of this debate is preoccupied by technical and operational questions. The questions it asks are: will conventional deterrence work?; can it do the job at acceptable cost?[18]

Close to the debate's surface, however, and explicit in many contributions to it, is the contention that conventional deterrence is desirable, not just for technical, but for political and ethical reasons as well.[19] This is most commonly refuted by the twin counter-claims that

---

[17] A. J. Pierre (ed.), *The Conventional Defense of Europe* (Council on Foreign Relations, 1986), 2.

[18] See e.g. J. Mearsheimer, *Conventional Deterrence* (Cornell University Press, 1983) and S. E. Miller (ed.), *Conventional Forces and American Defense Policy* (Princeton University Press, 1986).

[19] For one presentation of these arguments, in a particularly British context, see

conventional war is thereby made more likely and that our experience of twentieth-century conventional warfare does little to encourage faith in its conformity to just and limited war precepts.

Extended Deterrence

It may be wondered why it is necessary to treat the issue of extended deterrence as a separate question at all. Surely any assessment of extended deterrence derives in a straightforward manner from our assessment of deterrence itself: if it is acceptable to threaten nuclear war in self-defence, then it is presumably acceptable to do so likewise for the defence of one's allies, especially when the allies have taken so much of the initiative in securing this protection? And if deterrence is generally unacceptable, for whatever reason, it is no less so in a situation of extended threat than otherwise.

There is undoubted appeal in this logic. However, as the particular character of extended deterrence is seldom considered in the ethical debates about deterrence, it is worthwhile to pause and consider this situation. Are there any specific features of a policy of extended deterrence that may be considered significant to the theorist?

First, it has been contended that the system of extended deterrence spreads the risk of nuclear war geographically to territories that would not otherwise be targets of nuclear hostilities. For instance, it is commonplace in Australia to hear the objection that US facilities in the country simply make it a nuclear target which it would otherwise not be. Accordingly, in terms of limiting war in a geographical sense, it may be said that extended nuclear deterrence spreads security only at the risk of potentially greater insecurity to those countries that fall within the area of guarantee.

Secondly, it is frequently asserted that the system of nuclear deterrence is indivisible. The strength of the system, on this reasoning, derives from its cohesion, from its integrated operational capabilities, and from the knowledge that a challenge to a part is, by its very nature, a challenge to the whole. This seems to suggest two matters of ethical principle with practical implications. If the system is as integrated as this description suggests, might one not enjoy its benefits without dirty hands? That is to say that the indivisibility of the deterrent offers externalities to the free-rider who opts out of the system. As a case in

K. Booth, 'The Case for Non-nuclear Defence', in J. Roper (ed.), *The Future of British Defence Policy* (Gower, 1985).

point, Francophobes have often berated the 'independence' of Gaullist defence policy because France enjoyed the protection of NATO *de facto* even when its *de iure* links with the integrated military command were suspended. On the other hand, this very integration may, in more palpably moral terms, prevent an ally from opting out of the system altogether. States strategically vital to a wider alliance community might continue to enjoy the nuclear guarantee, even unwillingly, should they decide to pursue non-nuclear policies on their own. In other words, the very pervasiveness of the system of deterrence at once makes it practically easier for some, and in principle more difficult for others, to opt out.

Thirdly, and perhaps most importantly, if a chain is only as strong as its weakest link, the credibility of deterrence is weakest at that point where it is most extended. It is not surprising that deterrence theory has had to labour hard on the specific problems of restoring credibility to the US guarantee of Western Europe, at least as much to quell European doubts as those in the minds of the Soviet leadership. In this sense the credibility of the guarantee is only as strong as the general health of the alliance.

In part, the chosen instrument for restoring this credibility has been a strategy that would appear to carry least danger for the United States and hence seem a strategy that could be credibly executed. From this perspective the search for counterforce and other limited nuclear options has been driven, not by the need for discrimination, nor by the requirements of deterrence in general, but by the specific requisites of a strategy of extended deterrence.[20] Ironically, many of these endeavours, designed to give substance to the US guarantee, have themselves been divisive and have contributed to European unease and lack of assurance, either because they have made Washington appear even less likely to deliver upon her commitments, or frighteningly over eager to do so.[21]

However, various mechanical devices have periodically been sought to enhance this credibility. The search for 'coupling' has led to the development of policies that would make a US response more certain and automatic than might otherwise be believed. The stationing of US

[20] See e.g. A. H. Cordesman, 'Deterrence in the 1980s: American Strategic Forces and Extended Deterrence', in R. Nurick (ed.), *Nuclear Weapons and European Security* (Gower, 1984).
[21] M. Howard, 'Reassurance and Deterrence', in Howard, *The Causes of Wars* (Temple Smith, 1983).

personnel in Europe, with the risk of their loss in war, is a tangible expression of this quest. It has been supplemented by the deployment of US nuclear forces in Europe, most recently as part of the NATO INF modernization decision of 1979. It was argued that cruise missiles in Europe, under US control, would more likely be fired than US central strategic forces. Thereby European security would be more effectively coupled to that of the United States and, since the 'gap' at intermediate level would have been filled, the ladder of escalation could operate in reverse to prevent any form of hostility.

Whatever the merits or otherwise of the 1979 decisions, the point of interest for the present discussion is that extended deterrence, in its preoccupation with coupling, militates against geographical limitation. The logic again is to threaten a big war in order to stop a small one. Whether the risk is worth the benefit is not easy to determine but what can scarcely be gainsaid is that extended deterrence in time of peace constructs escalatory mechanisms which, if they are to be effective at all, would be very difficult to dismantle in time of war.

If there is a general conclusion to emerge from all this it might be the following. Given the logic of deterrence, it might be necessary for the *ius in pace* to be more permissive than the *ius in bello* since the latter is called into play in a situation of real hostilities. But we should not be so naïve as to assume that a liberal ethic of peace could be immediately transformed into a restrictive code of war should the need arise, not least because some of the structures of deterrence, such as its indivisibility and its coupling links, are specifically designed to prevent such a transition.

With this we are back to a central dilemma of nuclear strategic theory: what makes for good deterrence does not necessarily make for good execution in war. This perhaps counsels against the attempt to construct a code of behaviour appropriate to both. But neither is it clear that divorcing the discussion of deterrence from that of the actualities of fighting makes for either a sound military theory or for a palatable ethics relevant to the waging of war and peace.

CONCLUSION

IT is because the practice of war brings together the often competing realms of state action and individual judgement that the effort to comprehend it philosophically has the richness, and the difficulty, which it possesses. War, in recent human experience, has been an adjunct of the system of international relations in which the organization and means of states make demands upon individuals, but with the purpose, genuine or rationalized, of furthering individual values and needs.

In most cases the appeal to reason of state, which occurs with particular force in the 'necessitous' condition of war, is not the end of an argument but the beginning of one. Many of those realists who are sceptical of constraining state rights by appeal to cosmopolitan values are so, not because of self-indulgent delight in the brutalities of power politics for their own sake, but rather because they see the preservation of their particular state and possibly the international order in which it is located, as necessary for the protection of essential values. It is indicative of this that many of the post-war generation of realists were themselves *emigrés* from central Europe who had witnessed close to hand the cost in individual human values of an insidious form of political power not justly resisted. To this extent, the state justifies its wars by appeal to rights and liberties thereby guaranteed.

Unhappily, not all states function as the guarantors of rights: other delinquent states have consumed their own. It is because state boundaries serve ambiguously both to guarantee rights, but also to create their own internal 'killing fields', that the moral assessment of war is such a hazardous undertaking.

Philosophical analysis of war and the manner of its waging involves three distinct, but interrelated, phases. The first is a task of conceptual and methodological investigation—what is the nature of war and how, on that basis, should we proceed to think about the issues provoked by its conduct? Secondly, we are confronted by issues of substance—what is the most appropriate content of the rules we devise for the waging of war as regards such matters as weapons and targets? Finally, there remains the task of operational implementation—given the general principles developed for war, how can they effectively be translated to the battlefield itself?

The argument of this book has been that these are not separable tasks. The philosophical campaign to develop substantive rules of war has had to fight a war on two fronts, being pressed on one side by the demands of principle and logical consistency while being harassed on the exposed flank by the requirements of operability and the need to make philosophical theory meaningful in practice.

In the abstract these are mutually consistent tasks. It should be the objective of a practical philosophy to devise an understanding of war in terms of which certain principles of conduct would find their substantial theoretical bases. In turn, those general principles would yield tenets to be observed in the actual waging of war: the philosophical trinity of general concept, precepts, and codes of practice would thus be unified. The reality is, of course, by no means so straightforward. While the concept of war can contribute to the development of certain gross rules for its conduct, it tends to leave significant areas indeterminate: to say that war must conform to requirements of justice or political limitation sets but imperfect bounds upon its execution. Likewise, those purest philosophical principles for the waging of war are seldom ones which can apply immediately to the realities of war, as the earlier discussion of non-combatant immunity has suggested. At this point a further pragmatic choice must be made between adhering to the unadulterated principle or clinging instead to rules which lack philosophical weight but offer the advantage of greater ease of implementation: from this perspective salient categories may be more attractive than morally or politically compelling ones.

There are then difficulties at each of the philosophical stages of the argument. Most complex of all, however, is the interlocking nature of the problem inasmuch as the codes of war are torn between the demands of philosophical conviction and the requirements of practicability. Ironically, it is this philosophical praxis which creates the true unity of the field, bound together as much by internal tensions and contradictions as by consistency of philosophical principle and practice. All the more is this so as this is not a static realm of speculation: changes both in our ideas of warfare and in the technical characteristics of war contribute towards an ever-changing balance between the substance of the rules of war and the efficacy with which they can be implemented. Arguably, the major characteristic of contemporary warfare is the remoteness and physical separation between belligerents. This at once 'depersonalizes' the business of warfare and simultaneously makes enemy military forces the less obvious focus of our military endeavours. Traditional concepts of warfare are eroded by both these developments and a new synthesis is required combining the emerging theoretical and practical elements. Nothing has done more to hasten intimations of revolutionary change in warfare than nuclear technology. In August 1945, only days after becoming Britain's Prime Minister, Clement Attlee speculated on paper about the ramifications of the new weapon for British strategy. Having alluded to the changes in strategic practice which the weapon had wrought, Attlee continued: 'It is infinitely harder for people to realise that even the modern conception of war to which in my lifetime one has become accustomed is now completely out of date.'[1]

Attlee was manifestly correct in such an early appreciation of the impact of the atomic bomb on the conduct of war. Perhaps, however, he was wrong on two further counts. First—although this is with hindsight—conventional wars have persisted and, according to very approximate statistics, there have been some 150 of them since 1945, causing some twenty million casualties.[2] To this extent, traditional forms of war still account for much of the present reality of warfare. Secondly, and more significantly, Attlee's formulation may be misleading if accepted totally. If the suggestion is that, in reconciling ourselves to the nuclear age, we can afford to dispense with more traditional conceptions of warfare, it should be resisted. It is as palpably wrong to imagine that a single concept of war is equally valid for all historical experience as it is to believe that changes in the practice, or threatened practice, of war should immediately issue in the total abandonment of existing conceptual structures. To do so would be to reject all points of political and ethical reference in terms of which the new technology is to be judged. Contrary to the orthodox wisdom, there is a case for revolutionary technological change being met with *conservative* thinking, the more meaningfully thereby to tap existing traditions. This may sound like dangerous heresy but it must surely be preferable to confront the nuclear age with a set of existing ideas about the conduct of just and limited war than to face it in an intellectual vacuum. Were it otherwise, this book need not have been written.

The fact that warfare is a continuing human experience adds another element to the equation. How belligerents behave in war is affected by institutional memories of past wars and by anticipated

---

[1] Public Record Office, PREM 8/116, undated.
[2] L. Freedman, *Atlas of Global Strategy* (Macmillan, 1985), 51.

behaviour derived from past experience. This historical dimension should not be neglected. For instance, it forms part of the basis of Walzer's critique of the bombing of Hiroshima and Nagasaki and of the prevalence of hostage taking in various types of political strategy. It is his complaint that the negative consequences of past breaches of conventions can have long-term effects by contributing to more virulent styles of war in the future:

And if one is reckoning, what about all the future victims of a politics and warfare from which restraint has been banished? Given the state of our political and moral order, with which Hiroshima probably has something to do, aren't we all more likely to be victims than the beneficiaries of terrorist attacks?[3]

What this does is to provide a specific illustration of the dilemma of deducing rational behaviour in serial human situations: the sensible course of action in an isolated episode may not seem so if the situation is part of a sequence likely to be repeated. Game theory offers its own insight into this question: what constitutes rational behaviour in one game of chicken might not represent rational behaviour in a game that is to be serially replayed. As various analysts have suggested, in serial games, the factor of bargaining reputation can be of some importance in affecting the outcome of the game as those with good bargaining or bluffing reputations have impressive deterrent resources at their disposal. If rationality cannot be determined on the basis of a single episode, likewise judgements about warfare must comprehend the continuities within its practice and the influence, intended or otherwise, which past experience can have upon the future.

The relevance of this to a discussion of the conduct of war is twofold. First, and most obviously, how belligerents prepare themselves to wage war will reflect a combination of anticipations derived from future projections and past experiences. The calculation that an enemy will not abide by rules of warfare, on the basis of his reputation in war, is likely to shape the opponent's willingness to restrain his own war practices if reciprocity is not reasonably to be expected.

Secondly, and more profoundly, while it may be rational to infringe a war convention for reasons of military necessity or to save some of your own combatants' lives in a critical situation, it may well be that the cost of this is the loss of the life-preserving value of this convention in future wars. Accordingly, Fotion and Elfstrom discuss some problems

---

[3] 'An Exchange on Hiroshima', *New Republic*, 23 Sept. 1981, 14.

of the surrender convention in this context of the tension between saving lives and saving the convention (and possibly other lives) in the future.[4] The sequential nature of the experience of war, and the fact that future anticipations are related to past practices, entails the knowledge that decisions that appear to be cost-free on a short-term calculation are not so when regarded in a longer perspective.

The philosophical tools available for exploration of these issues are not entirely adequate to the task. The tradition of just war analysis is vulnerable to the twin charges that it offers, at best, a set of generalized precepts which are poorly adaptable to the realities of warfare and, in any case, seem to be doubtfully relevant to some of the major changes in the practice of warfare brought about in the present century. Even so, the revival of just war theory in the face of the challenges of total war and of the development of nuclear weapons seems eloquent testimony to the enduring philosophical need for such guidance and of the persistence of the ideological appeal which the doctrine holds. There are many writers who point to the shortcomings of the just war tradition but few who are willing to abandon this style of reflection altogether.

Likewise, the limited war tradition rests on insecure contemporary foundations. Its main doctrinal strength resides in its attempt to relate the means and ends of war in some politically intelligible fashion and to proscribe those means or levels of violence that go beyond any sensible political calculus. However, the doctrine remains open-ended in the age of infinitely expandable goals of policy, at a time when the concept of war itself cannot be expected to put a particular value on the goal of national survival. As long as this value remains indeterminate, and even in the face of instruments for the defence of national security which directly threaten that very goal, limited war tradition offers some procedural guidance but little by way of absolute standards. This weakness in theory is compounded by the challenge of practice in that limited war doctrine is parasitic upon a concept of escalation which, in turn, threatens to undermine the objective of limitation. As Thomas Schelling long since pointed out, it is the knowledge of potential escalation leading to the overturning of the boat that should induce caution in the two men who rock it, but the process of rocking it is itself open-ended and, leaving something to chance, capsize cannot be guaranteed against.[5]

[4] *Military Ethics* (Routledge & Kegan Paul, 1986), 145.
[5] See generally his *Arms and Influence* (Yale University Press, 1966).

Not only do contemporary traditions for the conduct of war have to cope with unprecedented political and technological challenges—with vertical proliferation of warfare—they have also to contend with war's colonization of new pieces of intellectual territory—with horizontal proliferation as well. Deterrence theory and arms control are two obvious instances of the imperialism which has taken place in the theory of war. The problems of *ius in pace* have already been reviewed and demonstrate the manner in which the traditional discussion of the proper conduct of war has expanded to encompass issues pertaining to the proper conduct of peace. The difficulty of separating out war from the workaday world of international politics means that the many instruments for the managing of international society—the politics of East–West relations generally, the legitimacy of intervention in the affairs of other states, as well as the more obvious issues of the arms race—have all become proper objects of concern for just war theorists. In a world of crisis management, in which the contestants communicate with each other by means of violent 'signals', it is all the more distorting to differentiate sharply between the language of war and the language of international diplomacy.

Similarly, the field of arms control has become much more closely integrated into the philosophy of war than traditional conceptions of disarmament. In the pre-1945 paradigm, arms were associated with the causing of wars and the only interest in disarmament was in reducing the incidence of violence which might flow from unbridled arms competition. There was little sense of arms control as the continuation of strategy by other means, far less a sense of arms control as an aspect of war itself and subject to the same philosophical and moral strictures.

As arms control theory has evolved away from disarmament towards a concept of strategic management, in which the balance and the nature of the weapons have been of more interest than absolute numbers, and certainly at the expense of concerns to reduce these numbers, so it has become intellectually caught up in deterrence theory and in the rehabilitation of peace by the preparation for war. The net effect of this has been for arms control itself to become enmeshed in traditional *ius in bello* questions: the arms control dialogue may proceed by way of technical disquisitions on launch ·vehicles, warhead numbers, accuracy, survivability, penetrability, and security of control but it is none the less informed by the same doctrinal concerns as the wider realm of deterrence theory and makes gross judgements about the kind of war-making capabilities that we should seek to maintain and those that we should seek to eliminate. Such peacetime discussion of force capabilities, and operational requirements, cannot sensibly be divorced from the actualities of the waging of war.

Perversely, what may appear to be a dismal and pessimistic development, namely the erosion of the distinction between peace and war and the more obvious intrusions of military infrastructures into the management of the peace, may not be without its more hopeful aspect. If the objective of the just war is the restoration of a better peace, ultimately the focus of attention must be upon the political, economic, social, and technological constituents of such a peace. If the pressures of the prospect of war concentrate human energies upon this task, at least some positive benefits might be derived from this gloomy situation. After all, the classical just war theorists argued for their own particular integration of peace and war and sought to guide the conduct of war in accordance with the principles of peace. In an age of endless cold war this becomes an everyday task and not one that can be delayed until the international society has finally collapsed in a state of war.

BIBLIOGRAPHY

ARON, R., *Clausewitz: Philosopher of War* (Routledge & Kegan Paul: London, 1983).
BAILEY, S., *Prohibitions and Restraints in War* (Oxford University Press: London, 1972).
—— *How Wars End* (2 volumes, Clarendon Press: Oxford, 1982).
BALL, D., and RICHELSON J. (eds.), *Strategic Nuclear Targeting* (Cornell University Press: Ithaca, 1986).
BAYLIS, J. *et al.*, *Contemporary Strategy* (Croom Helm: London, 1975).
BEITZ, C., *Political Theory and International Relations* (Princeton University Press: Princeton, 1979).
—— *et al.* (eds.), *International Ethics* (Princeton University Press: Princeton, 1985).
BEST, G., *Humanity in Warfare* (Weidenfeld & Nicolson: London, 1980).
BLAKE, N., and POLE, K. (eds.), *Dangers of Deterrence: Philosophers on Nuclear Strategy* (Routledge & Kegan Paul: London, 1983).
BOND, B., *War and Society in Europe 1870–1970* (Fontana: London, 1984).
BRACKEN, P., *The Command and Control of Nuclear Forces* (Yale University Press: New Haven, 1983).
BRODIE, B., *War and Politics* (Cassell: London, 1973).
BUCHAN, A., *War in Modern Society* (Collins: London, 1966).
BUTTERFIELD, H., and WIGHT, M. (eds.), *Diplomatic Investigations* (George Allen & Unwin: London, 1966).
CASSESE, A. (ed.), *The New Humanitarian Law of Armed Conflict* (Editoriale Scientifica: Naples, 1979).
CHILDRESS, J., *Moral Responsibility in Conflicts* (Louisiana State University Press: Baton Rouge, 1982).
CHURCHILL, R. P., 'Nuclear Arms as a Philosophical and Moral Issue', *Annals* (American Academy of Political and Social Sciences), September 1983.
CLARK, I., *Limited Nuclear War* (Princeton University Press: Princeton, 1982).
—— *Nuclear Past, Nuclear Present: Hiroshima, Nagasaki, and Contemporary Strategy* (Westview Press: Boulder, 1985).
CLAUDE, I., 'Just Wars: Doctrines and Institutions', *Political Science Quarterly*, Spring 1980.
CLAUSEWITZ, C. von, *On War* (trans. Howard, M., and Paret, P., Princeton University Press: Princeton, 1976).
COHEN, M., *et al.* (eds.), *War and Moral Responsibility* (Princeton University Press: Princeton, 1974).
COHEN, S. T., 'Whither the Neutron Bomb? A Moral Defense of Nuclear Radiation Weapons', *Parameters*, June 1981.

DANTO, A. C., 'On Moral Codes and Modern War', *Social Research*, Spring 1978.

DAVIDSON, D. L., *Nuclear Weapons and the American Churches* (Westview Press: Boulder, 1983).

DAVIS, H. (ed.), *Ethics and Defence* (Blackwell: Oxford, 1986).

DRAPER, G., 'The Idea of the Just War', *Listener*, 14 August 1958.

FISHER, D., *Morality and the Bomb* (St Martin's Press: New York, 1985).

FOTION, N., and ELFSTROM, G., *Military Ethics* (Routledge & Kegan Paul: London, 1986).

FREEDMAN, L., *The Evolution of Nuclear Strategy* (Macmillan: London, 1981).

—— *Atlas of Global Strategy* (Macmillan: London, 1985).

FULLER, J. F. C., *The Conduct of War, 1789–1961* (Methuen: London, 1961).

FUSSELL, P., 'Hiroshima: A Soldier's View', *New Republic*, 22 August 1981.

GALLIE, W. B., *Philosophers of Peace and War* (Cambridge University Press: Cambridge, 1978).

GIDDENS, A., *The Nation-State and Violence* (Polity Press: Cambridge, 1985).

GLOVER, M., *The Velvet Glove* (Hodder & Stoughton: London, 1982).

GOODWIN, G. (ed.), *Ethics and Nuclear Deterrence* (Croom Helm: London, 1982).

GRAY, C. S., 'War Fighting for Deterrence', *Journal of Strategic Studies*, March 1984.

GREENWOOD, C., 'The Relationship between *ius ad bellum* and *ius in bello*', *Review of International Studies*, October 1983.

HALE, J. R., *War and Society in Renaissance Europe* (Fontana: London, 1985).

HAMPSHIRE, S. (ed.), *Public and Private Morality* (Cambridge University Press: Cambridge, 1978).

HANDEL, M. (ed.), *Clausewitz and Modern Strategy* (Frank Cass: London, 1986).

HARDIN, R. (ed.), *Nuclear Deterrence: Ethics and Strategy* (University of Chicago Press: Chicago, 1985).

HARE, J., and JOYNT, C., *Ethics and International Affairs* (St Martin's Press: New York, 1982).

HOFFMANN, S., *Duties Beyond Borders* (Syracuse University Press: Syracuse, 1981).

HOWARD, M., *War in European History* (Oxford University Press: Oxford, 1976).

—— *War and the Liberal Conscience* (Temple Smith: London, 1978).

—— (ed.), *Restraints on War* (Oxford University Press: Oxford, 1979).

—— 'On Fighting a Nuclear War', *International Security*, Spring 1981.

—— *Clausewitz* (Oxford University Press: Oxford, 1983).

—— *The Causes of Wars* (Temple Smith: London, 1983).

JERVIS, R., *The Illogic of American Nuclear Strategy* (Cornell University Press: Ithaca, 1984).

JOHNSON, J. T., *Ideology, Reason and the Limitation of War* (Princeton University Press: Princeton, 1975).

—— *Just War Tradition and the Restraint of War* (Princeton University Press: Princeton, 1981).

—— *Can Modern War be Just?* (Yale University Press: New Haven, 1984).

JONES, C. D., 'Just Wars and Limited Wars: Restraints on the Use of Soviet Armed Forces', *World Politics*, October 1975.

KEEGAN, J., *The Face of Battle* (Jonathan Cape: London, 1976).

KEEN, M. H., *The Laws of War in the Late Middle Ages* (Routledge & Kegan Paul: London, 1965).

KENNY, A., *The Logic of Deterrence* (Firethorn Press: London, 1985).

LAWRENCE, P. K., 'Nuclear Strategy and Political Theory: A Critical Assessment', *Review of International Studies*, April 1985.

LIDER, J., *On the Nature of War* (Saxon House: Farnborough, 1977).

LINKLATER, A., *Men and Citizens in the Theory of International Relations* (Macmillan: London, 1982).

LUARD, E. (ed.), *The International Regulation of Civil Wars* (Thames & Hudson: London, 1972).

LUTTWAK, E., 'How to Think About Nuclear War', *Commentary*, August 1982.

McNAMARA, R. S., and BETHE, H., 'Reducing the Risk of Nuclear War', *Atlantic Monthly*, July 1985.

McNEILL, W. H., *The Pursuit of Power* (Blackwell: Oxford, 1983).

MANDELBAUM, M., *The Nuclear Future* (Cornell University Press: Ithaca, 1983).

MARTIN, L., *The Two Edged Sword* (Weidenfeld & Nicolson: London, 1982).

MEARSHEIMER, J., *Conventional Deterrence* (Cornell University Press: Ithaca, 1983).

MELZER, Y., *Concepts of Just War* (Sijthoff: Leyden, 1975).

MIDLARSKY, M. I., *On War: Political Violence in the International System* (Free Press: New York, 1975).

MILLER, S. E. (ed.), *Conventional Forces and American Defense Policy* (Princeton University Press: Princeton, 1986).

—— and EVERA, S. van (eds.), *The Star Wars Controversy* (Princeton University Press: Princeton, 1986).

NARDIN, T., *Law, Morality and the Relations of States* (Princeton University Press: Princeton, 1983).

NISBET, R., *The Social Philosophers* (Paladin: London, 1976).

NURICK, R. (ed.), *Nuclear Weapons and European Security* (Gower: London, 1984).

NYE, J., *Nuclear Ethics* (Free Press: New York, 1986).

—— 'Nuclear Winter and Policy Choices', *Survival*, March/April 1986.

OAKESHOTT, M., *On Human Conduct* (Oxford University Press: Oxford, 1975).

O'BRIEN, W. V., *The Conduct of Just and Limited War* (Praeger: New York, 1981).

Office of Technology Assessment, *Strategic Defenses* (Princeton University Press: Princeton, 1986).

OKIN, S. M., 'Taking the Bishops Seriously', *World Politics*, July 1984.

PARET, P. (ed.), *Makers of Modern Strategy* (Princeton University Press: Princeton, 1986).

PASKINS, B., and DOCKRILL, M., *The Ethics of War* (Duckworth: London, 1979).

PEARTON, M., *The Knowledgeable State* (Burnett Books: London, 1982).

PHILLIPS, R. L., *War and Justice* (University of Oklahoma Press: Norman, 1984).

PIERRE, A. J. (ed.), *The Conventional Defense of Europe* (Council on Foreign Relations: New York, 1986).

POTTER, R., *War and Moral Discourse* (John Knox Press: Richmond, 1973).

PRANGER, R., and LABRIE, R. (eds.), *Nuclear Strategy and National Security* (American Enterprise Institute for Public Policy Research: Washington, 1977).

RAMSEY, P., *War and the Christian Conscience* (Duke University Press: Durham, 1961).

—— *The Limits of Nuclear War* (Council on Religion and International Affairs: New York, 1963).

—— *The Just War* (Charles Scribner's Sons: New York, 1968).

RIZZO, R. F., 'Nuclear War: The Moral Dilemma', *Cross Currents*, Spring 1982.

ROBERTS, A., and GUELFF, R. (eds.), *Documents on the Laws of War* (Clarendon Press: Oxford, 1982).

ROPER, J. (ed.), *The Future of British Defence Policy* (Gower: London, 1985).

RUMBLE, G., *The Politics of Nuclear Defence* (Polity Press: Cambridge, 1985).

RUSSELL, F., *The Just War in the Middle Ages* (Cambridge University Press: Cambridge, 1975).

SCHELLING, T. C., *Arms and Influence* (Yale University Press: New Haven, 1966).

SIPRI, *The Law of War and Dubious Weapons* (Almquist & Wiksell: Stockholm, 1976).

SMITH, M., *British Air Strategy between the Wars* (Clarendon Press: Oxford, 1984).

STRACHAN, H., 'Deterrence Theory: The Problems of Continuity', *Journal of Strategic Studies*, December 1984.

TAYLOR, T., *Nuremberg and Vietnam* (Quadrangle: Chicago, 1970).

TILLY, C. (ed.), *The Formation of National States in Western Europe* (Princeton University Press: Princeton, 1975).

TOOKE, J. D., *The Just War in Aquinas and Grotius* (SPCK: London, 1965).

VINCENT, R. J., *Human Rights and International Relations* (Cambridge University Press: Cambridge, 1986).

WAKIN, M. (ed.), *War, Morality and the Military Profession* (Westview Press: Boulder, 1979).

WALTZ, K., *Man, the State and War* (Columbia University Press: New York, 1959).

WALZER, M., *The Revolution of the Saints* (Harvard University Press: Cambridge, 1965).

—— *Just and Unjust Wars* (Basic Books: New York, 1977).

—— 'An Exchange on Hiroshima', *New Republic*, 23 September 1981.

WASSERSTROM, R. (ed.), *War and Morality* (Wadsworth: Belmont, 1970).

WILKENING, D., *et al.*, 'Strategic Defences and First-Strike Stability', *Survival*, March/April 1987.

WINTERS, F., 'Ethics and Deterrence', *Survival*, July/August 1986.

WOHLSTETTER, A., 'Bishops, Statesmen, and other Strategists on the Bombing of Innocents', *Commentary*, June 1983.

INDEX

accreditation 12
'adversary partnership' 36, 60
allies, defence of 133
annexation 26
apartheid 49–50
Aquinas, Thomas 37–8
arms control 141–2
arms race 141
army: conscript 11; standing 15
Aron, R. 54, 56
atomic bomb 126
attack, wrongful 37
Attlee, Clement 138
Augustine of Hippo 33, 37

Bailey, S. 19 n. 18
Ball, D 102, 106 n. 19
bargaining process 69, 97
battle 6, 10; reach of 75; trial by 20
Baylis, J. 58 n. 9
Beales, A. C. F. 5 n. 9
behaviour, rational 139
Beitz, C. 5 n. 12
belligerency: collective 49–50;
  recognition of 83
belligerents: bargaining 61; behaviour of
  138–9; preparation of 139; right cause
  of 36; separation of 137–8
Best, G. 11, 47, 90 n. 21
Bethe, H. 103 n. 11, 108 n. 22
Bierzanek, R. 82 n. 6
bluffing reputations 139
Bodin 14
bombing: area 102, 129; strategic 1, 90,
  102
Bond, B. 90 n. 22
Booth, K. 120, 132 n. 19
Bracken, P. 56 n. 7
Brandt, Richard 127
Buchan, A. 21
Bull, H. 22–3
Bundy, McG. 130 n. 15
Buzzard, A. 104

Canterbury, Archbishop of 63
Cassese, A. 86 n. 18

casualties 26; reducing 77; rights of 91–2
Catholic Church: and deterrence 126;
  and proportionality 119
causes, just 37
'cheating' 20–1
Christian Church: disintegration of ideal
  40; fragmentation of 52; and just war
  33–4; see also Augustine of Hippo;
  Aquinas, Thomas; protection for 39;
  and resort to arms 33–4; and right
  intention 38; see also pacifism
Churchill, R. P. 105 n. 17
civil-military relations 56
civil war, 12, 16, 83
civilians: see non-combatants
Clark, I. 53 n. 1, 104 n. 13, 126
Claude, I. 46 n. 17
Clausewitz 53 ff., 61, 65–6, 68, 72; *On
  War* 54
codes of conduct 25, 66–7; in irregular
  war 84–5; of just war 36, 66–7, 138;
  problems with 71; warrior guild 38–9
codes of moderation 40
codes of war 74–5, 137; conduct of
  hostilities 41; for deterrence 122–3;
  implementation problems 82, 86, 97
Cohen, S. T. 113–14
Cold War 19
'collateral damage' 43
combatants 83, 89–90, 91–2; in guerrilla
  warfare 84, 85, 95; individual 17; rights
  of 93
commonwealth 14
communication, inter-war 97
Comte 21
'constructive engagement' 50
conventions, infringing 139–40
Cordesman, A. H. 134
cosmology, divine 20
counter-city strikes 107
counter-silo strikes 105
Counter Reformation 40
counterforce 102, 104–5; alternatives to
  104; defect in 104–5; and extended
  deterrence 134; 'moral' 104; strategic
  instability 105